FARM and RANCH MARKETING

Johnnie Rosenauer
Michael D. Hennessey
James M. Mullen

Prentice Hall
Upper Saddle River, New Jersey 075458

Library of Congress Cataloging-in-Publication Data

Rosenauer, Johnnie.
Farm and ranch marketing / Johnnie Rosenauer, Michael D. Hennessey, James M. Mullen.
p. cm.
Includes index.
ISBN 0-13-485590-6
1. Farms—Marketing. 2. Ranches—Marketing. 3. Real estate business. 4. Real estate agents. I. Hennessey, Michael D. II. Mullen, James M. III. Title.
HD1393.R59 1998
333.33'5'0888—dc21 97-2371
CIP

Acquisitions Editor: Elizabeth Sugg
Managing Editor: Mary Carnis
Director of Production: Bruce Johnson
Production Supervisor: Marc Bove
Production Editor: Inkwell Publishing Services
Cover Design: Darla Royal
Marketing Director: Danny Hoyt

Printed in the United States of America

10 9 8 7 6 5 4 3 2 1

ISBN 0-13-485590-6

PRENTICE-HALL INTERNATIONAL (UK) LIMITED, *London*
PRENTICE-HALL OF AUSTRALIA PTY. LIMITED, *Sydney*
PRENTICE-HALL CANADA INC., *Toronto*
PRENTICE-HALL HISPANOAMERICANA, S.A., *Mexico*
PRENTICE-HALL OF INDIA PRIVATE LIMITED, *New Delhi*
PRENTICE-HALL OF JAPAN, INC., *Tokyo*
PRENTICE-HALL OF SOUTHEAST ASIA PTE. LTD., *Singapore*
EDITORA PRENTICE-HALL DO BRASIL, LTDA., *Rio de Janeiro*

Contents

Foreword

The real estate industry is exciting, varied, and complex at times. Each facet has its elements of specialization—knowledge of local, state, and federal regulations for the residential practitioner; zoning, financing intricacies, expansion prospects for the commercial/industrial practitioner; and production capacity of land, the proximity of the market for that production, and water and mineral availability to the land for the farm and ranch practitioner. It is with the latter the authors are primarily concerned.

A more qualified, experienced, and dedicated group would be difficult, if not impossible, to assemble. Under the gentle yet firm hand of Dr. Johnnie Rosenauer this compilation of information, guidance, admonition, and encouragement is offered in the spirit of teaching interested parties, divulging techniques and pitfalls, and inspiring the highest standards of moral and ethical conduct.

Enjoy this book, benefit and prosper.

Joe M. Cullinan

Preface

One of the main motivating factors in the creation of this book was the lack of specific marketing-related information on farms and ranches available to the licensee or consumer. Some of what is covered here can be found in extensive material for other forms of real estate. In fact, there are numerous books available dealing with real estate marketing/brokerage and investing. The readers of this book are encouraged to explore some of those resources for general background information. The intent here is to provide some of the key aspects of real estate directly pertaining to farms and ranches. Some readers may prefer to use the term rural property. In this book, these terms will be used interchangeably.

To offer the reader some insight into the general layout, Part I deals with an overview of getting into and staying in rural property brokerage. Notice that this segment is broken down into two distinct aspects. *Entering* the field of rural real estate is not the same as *staying* in the field. The outlook of this book is long-term and not of a short duration perspective. Some sources for ongoing professional development plus ideas for broker selection are also included. Broker selection is an important decision and as such, should not be made on the sole basis of convenience or familiarity.

Part II begins with material on how to determine the "right" market area in which to establish a farm and ranch brokerage practice. We believe this is a most critical decision. Other portions of Part II will focus on certain marketing activities and strategies. This will include the "nitty gritty" of promotion, showing, certain contract issues, and movement toward consummation of the sale. In reality, these factors tend to blend together in many sales, and there may not be a true beginning or end. However, we have made an effort to put these factors together in what we hope is a logical sequence so that each aspect can be reviewed and considered on its sole merit.

Part III contains the rural property analysis portion of this book. The appraisal process and "number crunching" aspects of rural property are reviewed. While every reader will not become a farm or ranch appraiser, it is important to establish a sound foundation in these areas to better understand the industry. Whether the rural real estate practitioner is selling to or for others, or planning to invest personally, some degree of financial and valuation analysis skill is required to be effective.

Part IV is the ecological and land management portion of this text. Such critical issues as soils and water are addressed. Some of the tools available for use in land management are introduced. To help bring these ideas together, a case history is provided of an actual ranch that underwent some significant changes in order to accomplish a new owner's objectives.

As the material contained in this book is reviewed, it is important for the reader to keep in mind that no claim is made that this is an all-inclusive and comprehensive textbook. In some cases, it may fall very short of a particular situation that the reader or rural real estate practitioner may encounter. This nation's land is diverse, with many uses and limitations. Likewise, the customs and practices of one area can vary significantly from those of another. The same will be true of the laws and regulations regarding any particular tract of land.

The authors of this book have all had somewhat varied training and experience in agricultural real estate. In addition, the authors' specialties in the field of rural real estate are diverse. We hope that this will allow us to present the reader a broader viewpoint. However, the authors neither collectively nor individually claim to be knowledgeable in **all** aspects of rural real estate across the United States. The readers of this book are encouraged to take the information offered and adapt it to the geographic makeup, customs, laws, and, very importantly, the people of their particular area of this nation. Best wishes to all readers in their endeavors!

ACKNOWLEDGMENTS

The intent of this book can be broken down into four major objectives:

1. To introduce the farm and ranch brokerage newcomer (or possible entrant) to some strategies that will help make the transition easier.
2. To remind veteran broker/salespersons of some of the things they need to do or to offer a fresh perspective that will help them become more effective.
3. To partially fill a void in the literature available to consumers or licensees on the subject of rural real estate.
4. To "give back" a little bit to an industry that has been very good to the authors of this book.

If the first three items can be accomplished with success, then the fourth item will also be accomplished.

We want to express our appreciation to the nice folks at Prentice-Hall for their support and encouragement in this project. Our Editor, Elizabeth Sugg, and her staff have been most helpful.

Our thanks to Darla Royal, the talented artist who provided the illustration work for this book and the artwork found on the cover. And, special appreciation goes to Sharon Dreyer, our "in-house" editor and computer guru who helped transform our jumbled words into complete thoughts and sentences.

We are grateful to our reviewers for their input. Their valuable assistance helped us to forge a better product. We are especially thankful to Joseph M. Cullinan, AFLM, GRI for his Foreword. Mr. Cullinan is a wonderful role model for our industry.

On a professional level we dedicate this book to a special group of people. Our great land, in many ways, is a bread basket for the world. An incredibly valuable asset is entrusted to the stewards of the soil, the farmers and ranchers of this great nation. These people are often an overlooked segment of our society; however, they are the hardest working, most efficient, and greatest risk takers around. Every year, the farmers and ranchers of our nation produce a bountiful supply of food and fiber for our homes by their acts of faith and their various talents. A grateful salute is given to the farmers and ranchers who comprise this small, but critical, segment of our population.

There are certain individuals in our lives who have a special impact from a career perspective. To that extent, the authors wish to thank the following people:

- Rick Rothe, for encouraging professional growth and development.
- H. Irving Schweppe, Jr., M.D., a great boss, mentor, and friend.
- Martin A. Graf, for being an excellent example of how to be a friend and businessman.
- E. W. Howard, Jr. (deceased), whose teachings to a young boy have served as standards of conduct for an entire lifetime.

On a more personal level, we wish to dedicate this effort to our immediate families. Their confidence in us and their invaluable support are the cornerstone for whatever we are able to accomplish in life. To that extent, we offer our love and appreciation to:

- Susan and Timothy Hennessey.
- Martha Mullen.
- Danell and Jessica Rosenauer.

About the Authors

JOHNNIE L. ROSENAUER is a Professor of Real Estate at San Antonio College in San Antonio, Texas. In addition to his academic duties, Dr. Rosenauer has an active farm and ranch brokerage as well as a game management consulting business in south and central Texas. Dr. Rosenauer holds a BBA in Marketing and Real Estate from Southwest Texas State University, an MA in Management and Higher Education from the University of Texas at San Antonio, and an EdD in Adult Education from Texas A&M University. An active speaker on real estate topics, Dr. Rosenauer is also the author of numerous articles and several texts.

JIM MULLEN is a licensed real estate broker with an active wildlife management consulting business. He holds a BS in Wildlife and Fisheries Science and an MAg in Agriculture from Texas A&M University. When possible, Mr. Mullen combines the two careers by assisting buyers in locating suitable ranches and then providing technical expertise after the sale. He lives with his wife, Martha, in Castroville, Texas.

MIKE HENNESSEY was born and raised in San Antonio, Texas. He earned his BS Degree in Animal Science and an MBA in Finance from Texas A&M University. Mr. Hennessey began appraising real estate in 1983. He is a State Certified Appraiser and has earned the SRA and MAI designations of the Appraisal Institute. Mr. Hennessey has thirteen years' experience appraising rural properties in south Texas. He has instructed several college Real Estate Appraisal classes.

About Joe M. Cullinan, AFLM

Having entered real estate in 1973, Joe Cullinan has chaired the San Antonio Board of REALTORS Farm & Land Committee, initiated the successful marketing sessions of that board, and attained the GRI designation. He has served as a Texas Association of REALTORS Director from San Antonio, and has been active in the Texas Chapter #22 of the REALTORS Land Institute, holding the office of President in 1984. He has served the REALTORS Land Institute as a committee member, as a committee chairman, and on the Board of Governors. He has taught as an associate Professor at the University of New Mexico and in the GRI program for the Texas Association of REALTORS. He presently holds the coveted Accredited Farm & Land Member designation. He has been honored as the Texas RLI Farm & Land Broker of the year for 1983 and as Farm & Ranch Specialist of the years 1992 and 1993 by the San Antonio Board of REALTORS.

He is currently serving as a Director of the Texas Association of REALTORS, was elected Vice President of Region 7 of the Texas Association of REALTORS for 1995 and 1996, served on the Executive Board of the Texas Association of REALTORS, and has been reelected to the San Antonio Board of REALTORS Board of Directors and the Texas Association of REALTORS Board of Directors. He most recently was awarded the prestigious "Lifetime Distinguished Service Award," as well as Farm & Ranch Specialist of the year for 1996 by the San Antonio Board of REALTORS.

I
UNDERSTANDING THE CAREER

1

Developing a Career in Farm and Land Brokerage

GETTING STARTED IN A CAREER IN FARM AND LAND REAL ESTATE

The decision to enter into the real estate area of farm and land brokerage is one that should require considerable thought and planning. The intent of this section of the book is to help anyone going through the decision making process to be aware of some factors that they should consider. The goal here is to help practitioners become "successful by design rather than disappointed and discouraged by default." This section addresses some of the developmental aspects of preparing for a successful entry into the field of farm and land real estate. Later sections deal with specializing in particular areas within the farm and land brokerage practice.

PERSONAL ANALYSIS

An astute person goes through an investigation stage before making a career choice. Part of this process should include a personal analysis. Questions such as those listed below should represent only a starting point in this very important and highly individualized review. There are no truly "right" or "wrong" answers universally applicable, but there are definitely some things that are very worthy of consideration.

QUESTIONS TO CONSIDER

Why am I attracted to this career field? Often in recruiting and counseling of potential salespeople, it has been found that the motivation to enter the real estate field is based more upon the desire to leave another career than it is to participate in the field of real estate. Talk about jumping "out of the frying pan and into the fire"! The advice on this point is that virtually all the outstanding people currently operating in the overall field of real estate and in the specialty areas of farm and land brokerage became involved and grew to success because they truly wanted to do the things involved with their chosen career path.

One resource to review is the book by Robert L. Shook entitled *The Real Estate People: Top Salespersons, Brokers, and REALTORS Share the Secrets of Their Success* (Harper and Row Publishers). For this book, Mr. Shook interviewed some of the very best real estate practitioners in the United States. His conclusion and the advice of these tremendously successful role models was to either make a personal commitment to do those things necessary to succeed or stay out of the business.

That view brings out the second point. Potential newcomers to this field must ask themselves if they really know what kind of effort it takes to succeed in the business. This effort is divided into several facets. At the initial stage there can be a considerable amount of work to become prepared. In fact, nearly every single topic covered in this book can represent an area to study and learn well beyond what is presented here. The goal here is simply to introduce these topics, not finalize the knowledge base for practitioners. Certainly no one can know everything, but that is no excuse for failing to learn all that is possible in order to professionally serve the people practitioners work with and for. It is a fact that the more that is learned, the more the broker/agent is aware of how much there is to know. School is never out for the true professional in the real estate field, or in any other field for that matter. It is everyone's responsibility to obtain a sound knowledge base in the initial stages of career development. Once a sound knowledge base is established, astute practitioners will endeavor to expand the base in a consistent manner throughout their work lives.

On another level, there is the need to understand what tasks are entailed in the role of a farm and land professional. Longtime practitioners, trainers, and teachers in the field of real estate are still amazed at the perception of some potential entrants into the field. Yes, there are many times when it is an exciting and rewarding career. And, there are some practitioners who cannot envision a career without being involved in some aspect of farm and land brokerage. But do not be misled. Just because some aspects of the real estate field are very, very wonderful, there is the other side of the coin as well. Working hard on a transaction for months, and frequently stretching into one year or more, and then having the deal fall apart due to something outside of the practitioner's control is a very difficult and disappointing experience. Unfortunately, this experience is a common one. It takes a special type of resiliency to overcome such discouraging occurrences.

Frequently, potential newcomers say things like "I like people," or "I enjoy looking at pretty farm/ranch land." Liking people and being able to work with people are very different things. Certainly, it is a pleasure for many people to look upon good land being used properly. But, that is like "comparing apples and oranges." The role of any real estate broker/agent is that of someone who is in the business of people/problem solving. That does not mean that all of the people a practitioner deals with are going to be problems. Rather, it means that by its very nature, real estate and the transfer of title to real estate present many problems inherent in the process. Some of those problems require an extensive amount of research and effort to resolve. That translates into a considerable amount of time spent looking and searching into the "details" of the transaction. Such detail work is often tedious and exacting as well as very far removed from the "fun" part of the job. The unaware and/or misinformed candidate for entry into the field can be overwhelmed by the demands associated with the "true" nature of the tasks that go along with the role of a real estate practitioner. Some people find this type of work most distasteful to perform.

The newcomer is counseled to spend some time talking with and observing successful practitioners in action. Readers need to see what practitioners really are doing and then ask themselves if they are "ready, willing, and able" to do such tasks to the degree necessary for personal, family, and company success.

The previous question brings up another very important point. Does the newcomer have the support of family in this venture? Often, spouses and children are not involved in this investigation stage, and therefore, are unaware of the level of demands placed on a farm and land practitioner. The long and often erratic hours, which also include weekends and evenings, can place undue strains on relationships if understanding and encouragement are not present. There are enough distractions and obstacles to overcome even when this support does exist. It is a very sound idea to honestly and openly discuss the benefits/advantages and the costs/disadvantages with all family members old enough to comprehend the significance of such a career decision.

An important part of that discussion for most people will center on finances. The nature of farm and land sales dictates that a newcomer have some monetary reserves to cope with the delay between entry into the field and the first "payday." It is not unusual at all for a closing to be scheduled several months after buyer and seller have agreed to the terms and conditions of the deal. It is also quite common for it to take several months before the newcomer gets a buyer and seller "on paper." And, compounding this problem is the unpleasant fact that expenses begin before the newcomer's first day on the job and continue to accumulate whether an income is realized or not. It is important for someone reading this to understand that the authors are not trying to be pessimistic or trying to keep newcomers out of the farm and land industry by making these statements. In truth, one of the larger reasons for writing this book is to help others "do the job better and quicker" than the average person entering the real estate field. The primary reason for this is that very little support material regarding farm and ranch land transactions is available. Advising newcomers that a person needs to stockpile some savings prior to entering the field is merely a realistic statement. There is an old saying in real estate that states a good salesperson is a "hungry" salesperson. If a broker/agent defines hungry as motivated to learn, grow, and excel, it is a true statement. However, if that statement is interpreted to mean a broker/agent who is motivated by fear of not having sufficient funds to meet regular financial obligations, it is a misleading and potentially dangerous falsehood.

There are other factors to consider when making a personal investigation and analysis. Hopefully, the factors outlined here will provide a starting point in the effort. Once the newcomer is well studied and has a firm understanding of the details, the practitioner is ready to begin investigating and considering other aspects of the complicated field of farm and land real estate.

Prelicensing

On the simplest level, the intent of prelicensing course work is to meet the minimum requirements for obtaining a real estate license in the state where the prospective broker/agent hopes to practice. However, this chapter will present a somewhat broader point of view. One must certainly respect whatever specific courses a state licensing commission (this regulatory body is called by a variety of names across the United States) dictates a prospective broker/agent must take. Beyond that, however, prospective practitioners would be well advised to consider what they need to take in order to expedite their success in the field. This is somewhat of an individualized process that will depend heavily upon each person's past education, experience, and strengths. Another critical factor is what courses are available in the prospective practitioner's area. And, of course, cost and timing must also be considered. Assuming there are several choices available, here are some suggestions that are also applicable to continuing education, which is addressed in an upcoming section of this chapter.

Knowledge Base Needed to Meet Client Expectations

The old saying about the customer always being right is not quite true in some opinions. We very strongly believe that the customer may not always be right, but is most definitely important! Newcomers to farm and ranch sales will need to determine the more common informational needs of a buyer or seller of land in the marketplace where they specialize. Can some of this "need to know" material be incorporated into the educational process from day one? If so, that is a most effective and efficient way to help the new people "hit the ground running." For example, suppose a common concern of land in the practitioner's chosen marketplace deals with the issue of mineral rights. If a prelicensing course is available that will explain issues like the definition of minerals, royalties, executory rights, surface rights, etc. as defined in the area by custom or law, a wise person would consider taking the course over another course that may be easier or more convenient, but does not provide a sound knowledge base.

Courses that Strengthen Personal Weakness Areas

It is human nature to shy away from things that we don't understand or subjects we don't have an aptitude for or much interest in. Based upon what is discussed in this book and upon interviews with effective practitioners, prospective practitioners should have a feel for some of those things they will need for success even if they are not particularly competent in these areas. Rather than avoiding such areas, we advise that they embrace these things wholeheartedly in their prelicensing coursework. For example, they may have an innate fear of math. Logically speaking, the successful farm and land practitioner is going to have to do a good bit of "number crunching." This can be anything from calculating the internal rate of return for a particular development based upon projections, to approximating the yields of different crops on a given piece of farmland. It could be adjusting the price per acre on other area sales based upon the depreciable improvements to determine an asking price for a new listing, or figuring out the closing costs, down payment, and debt service when the practitioner receives an offer on that listing.

The point here is that practitioners will have to use math skills in order to become all they desire in this field. So whatever practitioners have a problem with or fear of, if they need it to succeed, then they must bolster it up early on. After all, there is no "crime" in not knowing something. The crime is in not finding out once we are aware of our limitations. Use the classroom as a place to learn and grow. It is a much safer environment in which to make mistakes than the outside world.

Courses that Strengthen Personal Preferences

No, this is not double-talk. There are so many aspects of farm and land brokerage that it is not possible to be an expert in all areas. Most of the successful people in the field have carefully "notched out" a segment of the marketplace for themselves based somewhat upon their own interests and abilities. There is some real benefit to being known among buyers and sellers and other practitioners as the "resident expert" on a particular type of land or geographic area. Everyone needs to start somewhere and if you are truly interested in developing specific skills or knowledge in a particular aspect, it makes sense to begin there.

For example, suppose your interest is toward analysis and evaluation. It may be wise to consider taking a class in farm and land appraisal as part of your coursework. This could well open a field of opportunity that matches your talents. We believe that people should strive for success in areas that complement their natural orientation. It is possible to maximize your strengths while concurrently correcting your weaknesses.

Format Issues

While this is not the place for a lesson on learning theories and teaching methods, it is appropriate to point out that **what** a person learns is greatly influenced by **how** it is presented. The adult student, even many years removed from a formal classroom setting, can still learn effectively. However, adults learn differently from their younger counterparts.

When considering the course options available, try to find out about some of the factors being presented. Some class schedules are very intense, lasting six to eight hours a day or three to four hours an evening every day of the week until the course is completed. Other courses will only take one night a week or maybe a couple of hours a day, but the course will last for several weeks. One type of course is not better than another, but one may be better suited for some people than others.

The ability to grasp and understand the material will vary among individuals. It is important that prospective practitioners do not set themselves up for failure by taking a class that is paced too fast or too slow for them. Other aspects to investigate include teaching techniques used, i.e., pure lectures versus audiovisual, group problem solving, etc. Attendance requirements, testing, and instructor qualifications should also be investigated. The prospective practitioner needs to be a smart shopper and check out all options prior to making a decision. Much like the topic of broker selection, which will be discussed later in this book, making sound choices regarding the prelicensing coursework will increase a person's chances of success and satisfaction.

2 Broker Selection

Another very important aspect of developing a career in farm and land brokerage centers on choosing a broker. There is no telling how many times the question has been asked, Who is the "best" broker? While there is considerable discussion of what factors are critical, little doubt exists that the proper broker can make a dramatic difference in an agent's personal growth and success. The points presented here are offered to stimulate thinking and hopefully cause interested people to begin their own evaluation. Keep in mind that there are any number of factors that must be balanced and prioritized in the selection of a broker. Few, if any, candidates will have all of the qualities that people are seeking and in exactly the combination they desire.

BROKER SPECIALTY/TARGET MARKET

As illustrated in this book, there are many different career facets within farm and land real estate. An individual salesperson or broker cannot realistically expect to specialize in all of the facets. Once a focal point has been zeroed in on, begin to determine which brokers are also active in that aspect of the business. A guess is that there will only be a few brokers effectively working in that area. The concept of specialty as intended here can mean several things. Specialty can be geographic, or it can also be the type of property, or even a particular group of buyers. You must determine where you best fit into the marketplace in order to clarify which brokers to consider.

If you have not yet decided upon a target market to specialize in, the other factors presented here may help you to do so. Some salespeople pick a broker and then a specialty. Just keep in mind that success in one field of real estate does not automatically translate into competence in another. For example, a most effective residential brokerage company is not, in many cases, going to operate with the same degree of ability in farm and ranch sales.

Experience has proven that it is much more preferable, all other things being equal, to work in a firm that generally specializes in one aspect of real estate. It is very easy to become distracted when much activity and discussion focuses on another area other than the one being practiced. This is especially true for salespeople who are just beginning their careers.

TRAINING AND SUPPORT

All newcomers to any profession need some guidance and support in the early stages. Unfortunately, not every company provides this essential support. Real estate is no exception. There are some firms dedicated to providing their people with the help they need to increase their chances of success. Others will view this responsibility with a much lower priority. For the experienced, already effective salesperson this is not so critical. But for the newcomer to real estate or even just to farm and land sales, this is a most significant factor. Most brokers will state this importance, but some will not follow through. Ask for specifics such as what topics will be covered, in what order, and in what manner. Do not read into this statement that the need for a large size office or company with a full-time trainer on staff is being espoused. Bigger is not always better, especially in the field of real estate. But, it should be emphasized how extremely important planned, direct experiences designed to get salespeople off to a competent and confident start are in this field. Training given on a "catch as catch can" basis or scheduled whenever nothing else is going on is dangerous, as well as ineffective. Choose a broker willing to give you the help and support you deserve and require.

BROKER REPUTATION AND EXPERIENCE

Balancing the support received is the issue of broker reputation and experience. Very little good can come from associating with a broker who the salesperson is ashamed to be seen with. Reputation is a two-way street. The broker needs to be proud of the salesperson's efforts since, in a great many ways, the agents are the company to the people they meet. What the broker or company stands for and has done in the past can help or hinder an agent's progress as well. Investigate the companies that you are seriously considering by seeking insight from as many sources as possible in the area and in the real estate community. Often, what people do not say will provide as much feedback as what they do say.

One fact to determine is the sales volume of the company. It is unlikely the firm will "open up the books," but some insight into overall number of units sold and gross sales dollars is a reasonable area of investigation. Notice the trends over the last few years. Any erratic changes should be questioned. Are the company's trends in line with overall area growth or decline? If there is much decline in the region or in the company, that may be something to consider seriously. Also, determine how the business is spread out over the sales staff. It would be unrealistic to have everyone doing the same volume. But a common occurrence to watch out for is the case of a very small portion of the salespeople doing all the selling and making all the money. If this is happening, it is important to determine if you will be given equal access to business opportunities that come to the office.

You should also find out what selling experience the potential broker has had. It is most important that a broker have at least some personal success in the sales field. This does not mean that the broker should have been the top salesperson for two decades in a row, but rather that the broker has attained at least average production in sales. This shows some credibility and also gives greater assurance that the broker will have some awareness of the challenges faced by salespeople.

The debate over whether or not a broker/manager should still be actively selling is a frequent and often heated one. There is merit to both points of view.

The skills and talents needed to be an owner/manager are somewhat different from those required to be a salesperson. And depending on the size of the office, carrying out those administrative duties can be a full-time job, and then some. However, being out in the marketplace keeps a manager current and involved. Experience reflects that most brokerage firms that specialize in farm and ranch property are not so large as to justify full-time, nonselling management. However, there are many commercial, development, and residential companies with active farm and land divisions that very much need a management team devoted to the task of management alone.

ENVIRONMENT

As these points and other important topics are discussed, the prospective salesperson should attempt to get a "feel" for the offices under consideration. Prospective salespeople are advised to spend some time interacting with the other salespeople in an attempt to determine if a comfort level can be achieved. Some offices have a very close-knit, almost familial sales staff. Others are more reserved toward one another. Some people will swear by one and swear at another while other individuals will reverse the outlook. The key point to remember here is the importance of determining in advance, if possible, how pleasantly and comfortably you can become involved with a firm and the people who staff it.

Be wary of any person, or company, who discourages comparison shopping. The good firms know they will stand up to the scrutiny, and they encourage people to shop around. Likewise, it is a very good idea to make several visits and spend some time in the office watching and observing the activities and interpersonal interactions. Certain individuals in an office can "make or break" a newcomer if allowed to intervene. An "older" agent who takes the newcomer under wing and helps to nurture a career is always fondly remembered. These experienced people are not threatened by newcomers and are more interested in helping a newcomer develop and become a true asset to the company and profession.

This type of person is a wonderful complement to management support. The other side of the coin is the nonsupportive agent who "second guesses" everything newcomers do. This can even include questioning their entrance into the field. Frequently, this type of agent will view a newcomer as a threat and will resent anyone joining the company or industry. Such individuals need to be recognized as soon as possible and avoided.

While a newcomer can never be absolutely certain of making the right choice of broker and office beforehand, these considerations will increase your chances. Broker selection is somewhat different from other interviewing or application experiences you may have had. Instead of hoping to make a good impression and wondering if you will get the job, we suggest that the newcomer is an equal partner in this selection process. You should do yourself a big favor and "hire yourself" the best broker you can possibly find. Remember that the goal is not just getting into the business, it is staying in the business, succeeding, and continuing to grow.

CONTINUING EDUCATION

Once a person enters into a career as diverse and challenging as farm and land brokerage, the realization soon sets in as to how little is actually known about the field. Experiences in the field will help offset this somewhat. However, you will need to continue with your education if you are to achieve all of your goals. Setting aside the required continuing education as dictated by individual state licensing laws, the beginner will have some freedom to pick and choose from a wide variety of broadening experiences. Only a few of the many possibilities available are reviewed here. Our purpose is to acquaint you with as many of the available options as possible.

LOCAL BOARDS OF REALTORS

Many of the area Boards have seminars of specialized interest to the farm and land community. These sessions may be on a regularly scheduled basis such as quarterly or even more frequently. Practitioners who are members of the Board plus invited guests from the surrounding areas may meet to hear a guest speaker on a relevant topic.

Sometimes these meetings combine a period during which participants can promote a listing or request help to find a particular piece of property. Much learning and sharing can occur at such events. Practitioners should determine quickly if this avenue is available, and then use it.

EXTENSION SERVICE SEMINARS AND SHORT COURSES

In a similar vein, the Agricultural Extension Service, or any other name used in a particular location, will offer the opportunity to learn more about related subjects. Often, actual tours of demonstration farms and ranches are included in the course or seminar. Many times the Agricultural Extension Service will have on staff some truly knowledgeable people who are very experienced in their field of expertise. Information and access to this level of talent helps the real estate professional to be more informed and competent.

LOCAL COLLEGES AND UNIVERSITIES

Given the likelihood that few, if any, people have come into this industry totally prepared, there may be some academic training and support needed to enhance performance. This could be a course in Earth Science in order to better understand the principles of geology or geography. It may be a course in computers to become more computer literate in that ever-growing field, or an accounting or financial analysis course. It might even be a class to learn more about livestock. Whatever the case may be, there could be a void in a practitioner's knowledge base that could be filled by a local school. The class may also count toward a degree or broker license requirements. Even if the course is not taken for credit, the information gathered could help the practitioner to perform at a higher level of competency. Certain states have established real estate centers in conjunction with a college or university. Some of the funding for these centers comes from licensing fees. These centers are terrific sources of information and support. Ask for the center's publications list to get an idea of what is available. A very fine example is the Real Estate Center at Texas A & M University located in College Station, Texas 77843.

SELF-STUDY

Some of the best learning takes place in a private, self-paced situation. By devoting time to self-study, you can accomplish a lot. One method is to subscribe to certain professional publications that can broaden a practitioner's views and awareness. Do not make the mistake of oversubscribing to the point that it is impossible to find the time to read everything coming into the office. Pinpoint those periodicals that will broaden your horizons and not needlessly overwhelm you. Once suitable journals or magazines are found, establish a library for background information. Since a practitioner cannot afford to buy all the fine publications in this field, the local public library is a resource that can be tapped for current information. If the local library is up to par, then once a quarter or more regularly, you should go there and spend some time reviewing the periodicals that are most beneficial to your individual needs. Personal discipline is needed for success in many aspects of life and self-study is simply another aspect that requires a considerable amount of persistence. Remember, if it costs time and money to become better informed and educated, consider the price of ignorance.

MENTORS/KEY RESOURCE PERSONNEL

Some of the greatest assets available to practitioners in our specialty area of real estate are the other people directly involved in supporting fields. It is important for practitioners to continue their own personal development by recognizing and seeking out individuals who can help them in their quest for more knowledge. Some of these contacts may be extremely talented in a specific area where only a broad scale of knowledge is necessary for the practitioner. For example, when a ranch is sold, some windmill or water well work may need to be done. The new owner may desire to clean out and restock a pond with fish, or have a new one designed and built. Some strip clearing of brush may also be an important activity. In all honesty, after many years in the field, even the "seasoned" practitioner may be only partially informed on such technical areas as these. But there is no need to become totally proficient yourself as long as you know people who are. It seems that the "smart" practitioner is the one who recognizes limitations and seeks appropriate counsel to make up for the limitations.

Remember, this association with other key people is not one-sided. The practitioner can be a referral and resource person as well, at the same time offering solid experience and expertise to the buyer. Everybody wins in this situation. Search out people who know more than you and then solicit their advice and counsel. You may be most pleasantly surprised at the information that can be gleaned from such relationships. In today's world and with the "information explosion" a reality, there is no excuse for not continuing to grow and develop. From the old timer who has farmed and ranched in the area for fifty plus years to the young, new county agent with all those "new-fangled" ideas, there are people out there to help you if you will simply take the initiative. Keep in mind that school is never out for the true professional. Brokers and agents must stretch themselves and continue to grow and expand their horizons. You may become a role model and mentor yourself. These are the ways of growth.

II
UNDERSTANDING THE MARKETING PROCESS

3 Who Buys Rural Property

THE RECREATIONAL BUYER

The factors to consider in selecting the "correct" piece of property to meet the needs of the full-time farmer or rancher will be covered in another area of this chapter. This group of people will most likely have much experience and knowledge to base decisions upon. Realistically, however, in many parts of the nation, that "user group" represents a fairly small, although very important part of the overall population attracted to rural property. In fact, in some markets it can be truthfully stated that farm and land brokers and agents do not normally sell **to** farmers and ranchers. Rather, they sell **for** them. The person desiring to invest in a piece of rural property for personal and family recreational purposes is a very large part of the target market in many areas. This group has been alternately praised and damned by the "real" farmers and ranchers: blessed when these "city folks" are acting as buyers and giving the rural seller more money than can be received from a neighbor; and then damned for driving the prices up beyond logical economic levels based upon productivity. But whether the agents swear by them or swear at them, in many sections of the United States, this group of buyers is extremely important since they help to keep the marketplace active. For purposes of clarity, those people who are buying a place in which to retire or live full-time while commuting to a town or city to work are not going to be discussed here. They are a different target group from the one being presented.

If a practitioner is going to work with the recreational target market, it is best to realize some of the differences between them and the people involved in agriculture on a full-time basis. As a whole, "typical" buyers have successful businesses or professions that allow them to be able to afford the purchase of a farm or ranch land. They are frequently very astute in business and financial dealings, and you should give them the healthy respect for their business acumen that is properly deserved. However, there is a need to balance this with the fact that business is sometimes conducted slightly differently in rural areas. One of the easiest ways to accomplish this task is to focus on the buyers' major motivation or perceived benefit in purchasing this land. It is important to remember that there are many reasons for investing in rural property. Even in this one area of recreational purposes there are many different motives. This book will address only a few of the more common motives that we have identified after many transactions through numerous years of working in the business.

HUNTING

One very common motive for the purchase of farm or ranch land in certain areas of the nation is to utilize the land for hunting purposes. In terms of perspective of importance, good hunting has ranked as the highest or second highest criteria for the last fifteen years in over ninety percent of south Texas ranch sales brokered or observed by the authors. Depending on the agent's personal orientation and philosophy, there may be nothing or a great deal wrong with hunting. This book is not the proper forum for a debate on the issue of hunting. It is the proper place to say that in some sections of the rural land marketplace, an astute investor or practitioner must have some insight into how to evaluate land through the eyes of a hunter.

In some areas of the United States there is an abundance of public land where virtually any and every properly licensed person can hunt. In other cases, a hunter must apply for and be granted a permit to hunt on public land. In some states there is a distinction made between a resident and a nonresident in terms of hunting privileges and licensing fees. In other areas there is virtually no public land. It is up to the agent to become familiar with the conditions to be found in a particular marketplace. Obviously, the situation most conducive to the salesperson or investor is the one in which most hunting must be done on privately owned land. While hunting from a recreational standpoint is being addressed, remember that hunting can also be a definite income generator as well. It comes as a shock to some landowners, both new and old, to discover that after an income analysis has been performed that reflects all expenses involved, hunting may be one of, if not the most profitable use of the property. There are many hunting enthusiasts who are willing to pay top dollar for a positive hunting experience.

Agents who find themselves attracted to or working in an area known for its hunting should keep some factors in mind. Graduate degrees can be obtained in wildlife management, so only a few of the basics will be highlighted here for you to consider. You must be willing to "learn the language" if you hope to succeed in this aspect of rural land brokerage. Terminology and how it is applied in the field is critical. Being able to differentiate between various forbs or to browse plant identifications is a proper starting point. Learning about the life cycles and nutritional needs of the various types of wildlife is necessary, as well as some insight into proper planting of supplemental "food plots" or programs using automatic feeders. Some elementary knowledge about the judging of game in the field will add to your credibility. Census counts and harvest data can be most helpful. Access to proper recordkeeping on any or all of the game management actions taken by current owners can enhance your ability to obtain a better sales price.

Another real benefit of such knowledge and experience can be your continued involvement with the property. By helping to carry out some of the aforementioned activities, you can generate additional fees. Frequently the recreational buyer does not have the time, interest, or knowledge to take care of the details necessary to accomplish the establishment and continuance of game sources for quality hunting.

Fortunately, you do not have to learn this all on your own. There may already be game biologists actively practicing in your area. Certainly you can consult with the county agent and local Farm Services Administration (FSA) office or Natural Resource Conservation Service (NRCS) office for guidance and contacts. Subscribing to certain game management magazines will definitely be useful. There are several periodicals that blend technical knowledge and laymen's language quite well. The *Texas Trophy Hunter's Journal,* published in San Antonio, Texas, is an example. Another example is the *North American Trophy Hunters* of Marietta, Georgia. In some segments of the buying public, the need to know about such topics is based upon a strong desire to develop proper game habitat strictly for the purpose of the enjoyment of watching the game. Keep in mind that there is a difference between being a hunter and being a killer. "Shooting" with a camera is still hunting in the minds of many recreational land owners.

One last factor to point out in the area of hunting is the exotic game aspect. Some parts of the country are very much involved with nonnative game, with a history of stocking exotic game on some ranches going back more than fifty years. In the segment of Texas known as the "Hill Country," so many exotic species exist on a free-ranging basis that the Parks and Wildlife Department has conceded it would be virtually impossible to eliminate the exotic game at this time. Some of these animals have different characteristics and habits that warrant special experience and knowledge prior to your attempting to work extensively with them. In spite of the need for experience and knowledge, there is some limited movement toward development of certain exotic species as commercial meat sources similar to beef cattle operations. While this has become common in parts of Europe, no one currently knows what impact this trend will have in the United States. Wildlife is such an important topic that we have included an entire chapter later in the book on the subject.

FISHING

A fellow was recently seen wearing a shirt at an outdoor gathering with the inscription shown in Fig. 3-1. This attitude is common and intense among a segment of the farm and land market. While some states like Mississippi and Louisiana have developed intensive commercial fish raising and harvesting businesses, the focus here again is on the recreational user. There are some privately owned "pay to fish" operations scattered throughout the U.S., but most fish development is for personal, friend, and family use.

Again, the task is to learn something about the language and the various life cycles. Stocking amounts and fertilization activities also become important to learn. The reader needs to keep in mind that marketplaces will vary according to type of fish and procedures, but some principles are generally applicable. For example, a common error is for a new owner to expect the privately owned body of water to produce lots of big fish and to do so fast. Many new owners cannot understand why one little tank or pond of one half-acre of surface area is incapable of producing fifteen-pound large-mouth bass, forty-pound channel catfish, and lots of two-pounders of both types for the kids to catch whenever they desire. These are unrealistic expectations for almost any small body of water, especially within the first year or two of creation.

Information on what to stock, when to stock, and in what quantities is commonly requested by landowners. There is guidance available from several sources. On a national scale, the best resource may be the American Fisheries Society with its publication

Figure 3-1

The worst day of fishing is far better
than the best day working.

center located in Bethesda, Maryland. This professional organization provides much research and technical data to its membership, which is then used in their practices. There are some organizations similar to the American Fisheries Society on a statewide or regional basis that can provide additional guidance. For example, the Texas Aquaculture Association, P.O. Box 2948, College Station, Texas 77841 and the Texas Fisheries Association, in conjunction with the Extension Service, provide valuable data on the creation and management of small impoundments of water. These associations can help any practitioner or owner to get started.

In other parts of this book the topics of soil and water are addressed from a slightly different perspective. These topics are nonetheless important to the segment of the population interested in fisheries. They are important first in the actual site selection for a new body of water. Frequently, buyers want a site that is physically pleasing, but not necessarily logical. This is when a firm understanding of slope and runoff and the ability to read topographical maps become critical. The ability to interpret terrain changes to calculate drainage amounts in terms of acreage and volume is critical in making decisions about where to locate new bodies of water. Sometimes there is not an adequate amount of drainage to justify building an impoundment of the size a potential owner desires. Of equal significance is determining the source of the runoff that will supply the new pond.

For example, if this watershed runs through a neighbor's productive farmland, is there a chance the neighbor can or will dam up or divert the water flow, leaving the new pond owner "high and dry"? Or will there be contaminants such as pesticides and herbicides that are necessary and legitimate for the adjoining crops, but lethal to the fish in the pond receiving the runoff? How about silting? The farmer does not want the good soil in the pond and neither does the pond owner, but can it be controlled? Then there are questions dealing with soil suitability. The set of standards used for ideal soil for farming or grassland for ranching is not automatically applicable to pond creation and management. For example, in large parts of the nation, bottom land along rivers, streams, or creeks is highly prized for its productivity. This same soil is frequently very undesirable for impoundments. This type of soil allows seepage of water back toward the "natural" body of water. Furthermore, when flooding occurs, dikes or diversions can be destroyed, with the result that the tank dams can be eroded and the impounded body of water lost. In addition, there is the problem of overflowing that will cause the managed fish population to become contaminated with trash fish of an undesirable species, size, or count.

There are many factors to be aware of and informed about. Thank goodness there is also help available. If a particular property is being considered because of its fishing potential, it is suggested that the practitioner seek competent advice from an educated professional with experience in the area. The county agent, Extension Service representative, or local NRCS or FSA Office are good places to begin. Tests such as sample soil coring may best be done by a land moving operator. Do not forget to seek the help of a fishery biologist in private practice when warranted. All these resources can be most helpful on issues from site selection to impoundment design and fingerling stockings of new structures. In existing bodies of water, population samplings may warrant total fish kill and restocking in a worst-case scenario. In other situations, guidance may be needed for selective restocking and harvesting. No one person can know everything there is to know about this subject, but everyone involved in the farm and land industry where fishing is a factor needs to become somewhat educated in order to do their jobs more efficiently.

THE PRODUCER

Hanging on one of the authors' office walls is a plaque that his sister gave him many years ago that states the following:

Requirements to Be a Rancher Today

1. One piece of worn-out land too poor to farm.
2. Two head of livestock, preferably one male and one female.
3. A nice pickup truck with power steering, air conditioner, and radio for listening to the local football games.
4. Two bales of old hay, one fence post, and a spool of barbed wire.

5. A gooseneck trailer.
6. An understanding spouse with a job in town.
7. An optimistic attitude that it will rain again someday.
8. Not a requirement, but of great help is to have a sympathetic banker.

Not everyone involved in ranching fits the above criteria, but it does offer one perspective. The full-time farming or ranching families of today are a small segment of the United States population. Different ways of keeping score yield somewhat varying numbers, but most of the research pegs the "country folks" at only two to three percent of Americans. What a sharp contrast from three generations ago, when approximately two-thirds of those living in the United States could be found on the family farm or ranch.

The mental image of a farmer in dirt-covered overalls with a piece of straw between his teeth is long-standing. Likewise, the idea of a rancher in jeans with a wide hat and tall boots (preferably with spurs) persists. But in fact, many of the people in agribusiness as producers of food and fiber do not fit these stereotypes. A more accurate description of today's United States' producer would reveal an older person (the average age of farmers is in the mid-sixties, according to Department of Agriculture research) with significant capital tied up in equipment, land, and operating expenses. The use of modern business tools is the norm rather than the exception. For the most part, this very important group of entrepreneurs has fields or pastures as a "manufacturing plant." Their product is a harvestable crop or animal unit. But a few ties do remain to the "old time" image. Producers today must still squeeze as much work and return from their days and dollars as possible.

When a licensee deals with a full-time operator interested in a piece of property, the focus is somewhat different than when dealing with other segments of the marketplace. In this case the consumer often knows much more about the land than does the salesperson. This is a time when the practitioner must acknowledge and respect superior knowledge. Unlike the recreational buyer whose income is mostly, if not completely, from other sources, this potential buyer will look at the land as a production unit. Topics like soil and water, yields, allotments, government programs, etc. take on a very different meaning. While a "big city refugee" often thinks that draws and other changes in elevation add character to a tract of land, the producer will look at them in terms of erosion potential.

To help balance out the different perspective and strong knowledge of such a buyer, the licensee would do well to seek as much technical help as is available. This is when the Farm Services Agency, Natural Resource Conservation Service, Agricultural Extension Agent, and whoever else can be found come into the picture in a big way. Even if the producer does not agree with these technicians' viewpoints, at least it is not the licensee's ideas on the line! Another good idea for working with this group of buyers that has been effectively utilized is the "peer mentor." This term refers to a respected producer in the same geographic area or in the same specialty field of agriculture. For example, when Bobby Darrell, a well-known rancher in western Oklahoma, speaks or writes of embryo transplant techniques in his cattle operation, registered cattle breeder Sammy Paul, located in Montana, might find some insight in the comments. Hence the need to subscribe to several professional publications that deal with issues relevant to the licensee's marketplace. This allows the licensee to be somewhat current on topics and trends, although the licensee may not be personally involved in farming or ranching.

Another logical offshoot of this peer resource strategy involves the use of a mentor for the licensee. This person may be retired from farming or ranching or still going full blast. In either case the mentor can be a great help to a beginning or even veteran real estate person. A key point in identifying a mentor is that the individual should not feel threatened by being a resource for others. And now a word of caution: Do not quote the "center of influence" about anything or anybody without the person's knowledge and permission. This remains true whether the comments are positive or negative. Such insensitivity can quickly cut off a true lifeline of insight and business for the licensee. One more suggestion for the licensee regarding the use of mentors is to make sure they know of your appreciation for what they do for you and your business. Keep them informed of how things are progressing in the work they helped you accomplish. You should be certain, if you learn of anything that might help your mentor's resources, to pass along the information. A sample letter to a mentor is shown in Fig. 3-2 to illustrate this point.

This in no way suggests illegal payoffs in money terms. It is rather an attempt to describe an extremely valuable marketing tool available to the astute salesperson. When dealing with a group of people who are far and away more experienced in their business than you, it is best to seek assistance from as many sources as possible. As a result of the previously mentioned "graying" of farmers and ranchers, there is a tremendous body of experience that can be tapped into. Select your resources carefully and cultivate long-term relationships.

No matter what resources are available to you as a licensee, you should expect the producer to view a piece of land very pragmatically. To most buyers in this category, a piece of land is only pretty if it is profitable. Therefore, you need to be very certain of your facts on past crop performances, carrying capacities, etc. One veteran broker aptly stated this point when speaking about his western-state cattle country buyers. He said, "When you go to blowing smoke with these boys about what a piece of dirt will do, all you're gonna end up with is a burned deal."

Our collective experience in working with producers as buyers and sellers is that they are fair, but tough, negotiators. The majority of producers are dealing with ever-shrinking profit margins and are compelled to get the most productive property possible with their limited land investment dollars. This group is a critical buying segment in many areas, the only real "game in town" to sell for or to in some cases. In any one locale, their "circle" is fairly small. Deal unfairly with one and you could be ruined in that marketplace. Treat them fairly and honestly, and you could build one sale on top of another to this market group.

THE DEVELOPER

A very different population of buyers is represented in the group we refer to as developers. Without being too technical or specific,

Figure 3-2

Sample Letter to a Mentor

Mr. Ned Royal
1931 Bob Duncan Road
Pleasanton, Texas 78064

Dear Mr. Royal:

Thank you so very much for spending some time with me and Ray Bean the other day. Mr. Bean is seriously considering the information you shared with him regarding converting side row irrigation into pivots. Such insight can best come from people like yourself who have lived through the change on a farm and know firsthand the good and bad points of each system. It will be helpful if Mr. Bean decides to purchase the Rackley farm or another one in the area. As you suggested, I have introduced Mr. Bean to Mr. Cliff Richter to discuss the specifics of the water well that Mr. Richter drilled for Mr. Rackley.

I thought you might find the enclosed article about the revised peanut program helpful. I will keep you informed of anything else I might learn on the farm bill. Thank you again for being such a great resource for me and others. I truly appreciate your willingness to help.

Sincerely,

Stetson Aaron, REALTOR
The Levi Aaron Realty Company

this term is used to define buyers whose primary purpose in acquiring rural property is for resale. This broad group can be further broken down into classes. Examples include: the subdivider (buy it by the section and sell it by the acre), the builder (buy unimproved land, improve it, and then sell it), or the investor (buy land where future demand exceeds present demand). There are likely other labels, but hopefully the point is clear that these people are not interested in land for personal use or enjoyment. Rather, their acquisition of rural property is motivated by the profit potential it holds upon resale.

Like other potential buyers, developers desire accurate technical data. However, their "slant" on the land is somewhat different. Rather than worrying if the land will yield enough to raise "a crop or a calf," they will more likely want to know if the soil is stable enough to pour a slab foundation. Instead of concern with game populations, they will look for data on future people population trends. While an irrigation farmer may detest a sloping area full of mature trees, such a setting may excite the future "ranchette" seller. Someone desiring a personal "runaway" may value isolation and dislike a great deal of road frontage, but usually a developer is just the opposite.

Developers are sometimes criticized for "ruining" the charm of an area by buying up large tracts and selling the tracts off piece by piece. In this book we will neither defend nor denounce such actions. Whatever the licensee's personal feelings about the development process, the reality is that investors with such motivations are viable purchasers in certain rural marketplaces. Working with this set of buyers requires that you become well versed in urban-related issues. Topics such as the paving of roads or making the existing roads wider, the existence of or provisions for utilities, and the adequacy of potable water are all near and dear to the hearts (and pocketbooks) of developers. You may need to become very proficient in subdivision rules and regulations in your marketplace. Public limitations of use vary widely, sometimes even from county to adjoining county. Much depends on the outlook of the various local public officials. Whether you like it or not, you must become aware of the political situation and mind-set about development in your marketplace.

It is important to spend some time investigating population trends. On a federal level, the Department of Census is a starting point. But a more local perspective could include the Chamber of Commerce or an Economic Development Agency, if one exists in your area. Sometimes data is available on population work that has been done by one of the state universities. The more information you can provide to help this group of buyers see how they can make a profit, the more likely they are to buy. Please keep in mind that these buyers are going into the deal with the intent of selling again. Learn to evaluate a piece of land "through their eyes" and you may just find a new source of repeat business.

THE NEWCOMER

The last group of rural property buyers to be addressed here is the group called the newcomers. These are people who desire to move into a rural setting. There is a very interesting growth trend occurring among this market group. Attempts to determine why these people are moving out of the cities and suburbs have uncovered many motivations within this group. The reasons include fear for personal or family safety, being tired of congestion, perceived decline in public services, high property values and ever higher property taxes, and simply a strong desire for change. In fact, some of the strongest population growth areas in several states owe their growth to this market segment. Examples like Branson, Missouri or Kerrville, Texas come to mind. Keep in mind that builders and developers are acutely aware of these trends, also.

Some of the newcomers are in reality "returnees." They are from small town or country families and began life in a place similar to what they are now searching for. Maybe they are literally returning back home to where one spouse or even both grew up. Maybe they are retired and ready to "kick back" in the country. On the other hand, it is possible that they are now parents and desire to have their children grow up away from the big city and its problems. However, many of this group are true "tenderfoot types" with no more experience in a rural setting than a few childhood trips to their grandparent's farm. Keep in mind that just two generations ago, many people had immediate family still out on the farm. Whatever the motivation, there are numerous pockets of rural America where there is a high demand for property from this group.

In dealing with this marketplace, one common characteristic keeps recurring. It is the need to balance expectations with reality. The ideal country setting probably does exist somewhere, the one with productive soil, magnificent trees, abundant water, excellent neighbors, quaint yet modern improvements, and an isolated but still very accessible location. All this and at a modest price topped off with liberal owner financing. In our fifty-plus collective years of experience with rural real estate work, that property has not been found yet. But, should such a property be discovered, rest assured that every rural broker and salesperson will argue over who gets it!

Helping the newcomer adjust to what it will mean to live in a rural setting is sometimes a real challenge. The movie *Baby Boom* is familiar to most people. In the movie, Diane Keaton portrays a Manhattan executive who inherits an infant when some family members are killed in a plane crash. Ms. Keaton moves to a farm in Vermont and struggles with the changes in her life. The story is worth watching purely as an example of what some people will experience if they make such a move. Perhaps all rural property brokers should buy a copy of this movie and have all potential buyers view it **before** they request assistance in buying a place.

Another example of this change bears mentioning. A retired couple sold their home in a large city and bought a small farm. The nearest small town was about thirty minutes away. After settling in, they invited some long-time friends and neighbors from the city to spend a weekend. After one meal, the hostess, an excellent cook, served cake for dessert. The guests commented that after all their years of friendship and shared meals, they could not recall ever having been served this particular cake before. The hostess commented that it was a "Make Do" cake, one of many new make do recipes she had learned since moving. "I call it make do," she said, "because when it takes an hour to go to the grocery store, you just learn to make do with what you have."

Such a situation may seem most unappealing to some people. Often the person doubtful about such a drastic lifestyle change is a spouse or a child still at home. Licensees are advised to be careful in selling the rural lifestyle too hard. You may personally love such a situation, but just as in working with a developer or hunter, you should not let your personal bias affect your professional judgment. Rural life, or even life in a small community, is not for everyone. (To that, some rural folks will say, "Thank goodness.") Even for individuals who want to come, the adjustments can be difficult. Experience has proven that one of the greatest changes for the large city dweller is the lack of options. Less choice in cafes or doctors, hairdressers or mechanics, grocery stores or electricians—less is the norm in rural life. However, there is also the chance to really know one's neighbors, to have a child's talents and abilities recognized in school and appreciated, and to truly feel a part of the community. But there is nothing gained without something lost. The adjustment is just too great for some people. We advise that you present the facts about what rural life entails in your marketplace and allow the buyers to make their own decisions. After all, what is "right" for one person or family is not the same for another. You need to simply supply the canvas and colors; let the buyers paint their own picture.

Of all the groups that a practitioner may deal with in rural real estate transactions, the newcomer subset may require the most assistance and support. It is very important to work within the boundaries of proper disclosure at all times, but it is never more important than when dealing with the uninformed and inexperienced. Another chapter of this text addresses the issues of agency and disclosure. Regarding buyers of rural real estate, keep in mind that the classifications presented here are only for illustrative purposes. You may have an entirely different, unique section of the marketplace to work with. If so, we advise that you learn what is important to your buyers and sellers and provide accurate information in a way that helps them make informed decisions.

4 Choosing the Correct Market Area

FINDING THE "RIGHT" AREA

Just as there are numerous factors involved in determining the "right" geographic area for an investor, there are just as many considerations for a practitioner to consider before deciding in which locale to establish a practice. Surprisingly, many of the same items are found in both situations. The issues addressed here may not be one hundred percent complete, but there are enough criteria to provide a very good starting point for both the broker/agent and the buyer.

The concept of target market segmentation is not anything new or original to this publication. The idea is simply that no one person or product can be all things to all people. The idea of "homing in" on a particular area in which to specialize is not only smart, but vital for most businesses or businesspeople to survive. We believe that preparation of an area analysis should be the first step in selecting the proper marketplace for the broker/agent. What exactly is the practitioner looking for? Thank you for asking!

AREA ANALYSIS

One of the first important factors is that of **compatibility.** For potential buyers this means having to look for property that is compatible with their investment goals. It is amazing how often this critical factor is overlooked. An area rancher who wants to expand the size of his holdings and operation is going to have a different set of standards from a rancher turned developer who speculates on a tract by cutting it up and selling "ranchettes" to folks from the big city. The recreational buyer who seeks solitude from the high stress and pressures of the metropolitan workplace has an entirely different set of needs than does the farmer who is being displaced by urban growth. Both buyers may be "victims" of the city, but that is where the similarities end. Before you waste valuable time and energy, as well as the time and talent of others, you should attempt to clarify as specifically as possible what the expectations are, both short- and long-term, for this mythical perfect property.

For the practitioner, compatibility really addresses how comfortable you are within a particular marketplace. A successful broker/agent in a high dollar resort area like Aspen may suffer more than just culture shock if relocated to the wheat fields of Kansas. And, the practitioner who knows all about peanut farms in Georgia might find things different in the rimrock country of northern Arizona and New Mexico. None of these areas are better or worse than any other, just very different. Astute practitioners take the time to research where they are most comfortable.

Another important aspect to consider is **accessibility.** To the small investor with less than $10,000 to use as a down payment, an apartment building near New York's Central Park is just as impossible to own (at the present time) as is a one thousand acre highly improved horse farm in Kentucky. Knowing your limitations translates into working within your means to afford the investment. It also refers to sticking to a geographic limitation that makes good sense. There is a picture of an old cow standing up to her knees in good grass and leaning way over a fence to eat. Finding an appropriate investment in land does not always entail searching far and wide. The grass is not always greener across the fence.

For the practitioner, this can mean several things. One of the important points to consider in dealing with farm and ranch sales is the travel factor. A terrific speaker with a long history of success in the field once asked at a farm and land seminar how many of the licensees attending would gladly take a listing within one hundred miles of their offices in any direction if the listing was attractive and well priced. About eighty percent of the attendees quickly indicated they would take such a listing. The speaker's explanation shocked most of the experienced practitioners in attendance, and it is an important lesson to share. The speaker pointed out that the minimum market area as geographically defined was a territory larger than the state of Georgia, the largest of all states east of the Mississippi River! The harsh reality of farm and ranch sales is that broker/agents will most likely spend a considerable amount of time and money on travel in and around their market area. When selecting a trade area, find one that makes sense distance-wise.

Still another factor to be considered by both the investor and the practitioner is **competition.** It is not realistic or even advantageous for a broker/agent to be the only individual in the marketplace. However, if there are too many other practitioners who are effective in acquiring property or in brokering sales, the broker/agent may want to seriously consider that fact. Do not interpret this consideration as running away from competition. Quite the contrary. As an investor, if you can become involved in an area where other experienced, successful investors have acquired land, it might be a very good thing. An applicable example is searching for a place to eat when you are in an unfamiliar city. If the "golden arches" of McDonald's can be located, rest assured that in most cases, there will be several other food sellers located nearby. Other site selections will be based upon the market research performed by another firm. It is vital that you understand this factor: It is not wise to become a secondary purchaser **from** these astute investors—that is, unless you feel the sales price will still meet your investment goals with your profit and expense added on to the sales price of the property. Timing is very important when looking at what the competition is doing.

For the licensee, competition must also be seriously considered. In most marketplaces there will only be a small number of firms (usually fewer than a half dozen, based on our experience, and frequently fewer than that) that are providing exceptional service in a particular trade area. There may not be a single firm doing so. Evaluate who you will have to be working with (and sometimes competing against) for business opportunities. Generally speaking, it takes several years (and sometimes many years) for a person or company to establish a reputation sufficient to dominate an area. Likewise, it may take a very long time to overcome such control by another firm. The astute practitioner will give much thought to these factors before making a final decision on a marketplace.

RECEPTION

Sometimes the decision becomes easier to make based upon how well a practitioner is received in an area. A third or fourth generation native daughter or son of a family with a strong and positive local reputation is fortunate in being able to combine a lifetime of experience with the people and land in the area; they may have a real edge in getting started in local brokerage. Unfortunately, few practitioners find themselves in that situation. In order to attain success in a particular location, it is vital to be able to gain acceptance by the buyers and sellers that the practitioner will work with. A worthy example is that of one young woman and her success. As a former nurse, she concentrates on working with physicians who are looking for farm and land investments. By her own admission, she is a city girl who had no farm or ranch background prior to beginning her real estate career. Through hard work and persistence, she has created her own niche in the marketplace. Now, many sellers and brokers with new listings call her to ask if she has anyone in mind for the type of property they are trying to sell. On the other side of the sale, there is a retired high school teacher who grew up in farm and cattle country and then taught agriculture locally for many years. He knows and is known by everyone involved in cattle or farming in two or three counties. Consequently, when someone decides to sell the old family place, they call him first. If the new listing cannot be sold to a local, he will let the "big city" brokers, located about one hour away, cobroker the sale to someone from the city. Either way, the gentleman controls and gets credit for the sale.

These two examples are at the extreme ends of the spectrum in terms of how each practitioner arrived at finding a locale. Both practitioners achieved fine results for their efforts. The main concept here is to become accepted by those you will serve. A practitioner does not have to be exactly like the two examples in your chosen marketplace, but you do have to be welcomed into the area.

ACTIVITY

Elsewhere in this book, the issue of determining the value of a piece of land is addressed. In this section we only emphasize the importance of choosing a marketplace with adequate activity to meet the practitioner's investment or practice goals. The investor wants to look in an area where supply and demand are somewhat in balance. If too much demand exists, a seller can ask for and obtain artificially high prices. This can occur when a new business development is coming to an area where speculation exists. A valuable lesson for the seller in such a circumstance comes to mind. A very large entertainment facility was rumored to be coming to the edge of a large city in the Southwest. Much speculation and jockeying transpired as to which site would ultimately be selected. Then the announcement was made that the facility would definitely be built in the city, but the site had not yet been decided. However, it was between the southeast part of the city and the northwest edge. Strategically located properties in both areas immediately showed large increases in values. One family, landowners of a hundred acres of worn-out, infertile, dry cropland, was offered approximately twelve times the agricultural worth of their land. The family became greedy and countered at approximately fifteen times the production value. Two weeks later, "the other side" of town won the site selection and the greedy family's land dropped back down to its original value. It is worth remembering that "pigs get fat and hogs get slaughtered" when it comes to holding out for the last dollar.

Likewise, an investor does not usually desire an area where supply exceeds demand unless the prices are so low or property so exceptional that it can be justifiably held until market conditions change. Further, it just may take a very long period of time to market even good, well-priced places in a "buyer's market." That period of time can "eat up" a great deal of any profit potential that exists, and can make properties unattractive risks to many people. Remember that low price is not the issue to focus upon. Rather, it is the relationship between purchase price, improvement/production/holding costs, and final sales amount that makes a difference.

To the practitioner, a marketplace in some kind of balance is important for the same reasons plus a few more. If the demand is far greater than the supply, sellers will feel they have no need for a practitioner's help. This can be erroneous thinking because the seller may not be a skilled real estate marketer or negotiator. There is, in most cases, more to selling a tract of land than placing an ad, putting up a sign, and waiting for the buyers to flock into the office. Another factor is that when supply is short, sellers often place unrealistic price expectations on a broker. Overpriced properties frequently lead to poor reputations for both the tract of land and the listing broker. Practitioners know that they are in trouble with the price on a listing when their peers greet the announcement of a new listing with the question: "And how much overpriced is **this** one?"

If there is an excessive supply of properties for the demand, another set of troubles occurs for the practitioner. Buyers are quick to become acutely aware of this situation and in nearly every case, some will "go for the throat" when making an offer. Distress buyers and bargain hunters are sometimes dignified with the label "investor," but that may well be a misnomer. It is often simply a case of people wanting to take advantage of the misfortune of sellers. This situation will make a difficult marketing problem worse because the seller will become insulted by such low offers. Depending on who the practitioner represents in the transaction, this approach is appropriate or distasteful. In either case, an out of balance marketplace is problematic. For this reason alone, you should give serious thought to and conduct a thorough review of the history of any area that you are thinking of entering.

There may be other issues to consider when determining the correct marketplace. As these issues are considered, keep in mind that there is a constant ebb and flow of products and uses of the land in the United States. Certain farm commodities and development areas that are valued far below production costs at one point in time become scarce and highly prized during another period. Knowing **when, what,** and **where** is one of the ingredients to success in any venture. With many variables to consider, landowners frequently try out new crops and uses of their land in order to strike a decent balance between supply and demand. The reader's role may be that of an investor or a broker/agent (or maybe both). In either case, carefully analyzing areas before committing to any one marketplace will impact your chance of success as much as any other item.

5 Marketing Farms and Ranches

THE ACTIVITIES AND STRATEGIES NECESSARY FOR SUCCESS

The Mythical Easy Sale. It is certainly true that some properties will virtually sell themselves. And, there are some sales that are so simple that most anyone could see the transaction through to completion. However, keep in mind that such transactions are the definite exceptions. Effective marketing of real estate, any type of real estate, takes a combination of skill, effort, experience, persistence, and sometimes a healthy dose of good fortune. While everyone appreciates a generous stroke of good luck, the reality is that luck alone will not be sufficient to get the job completed in the vast majority of cases. This chapter discusses some of the fundamental marketing activities required to obtain and then successfully sell farm and ranch land on a consistent basis. Rather than just telling practitioners **what** to do, this chapter will also include the strategies behind these activities. We believe that average people know what to do, but superior people know **why** they are doing those things.

While our main focus in this particular chapter is on the practitioner, the investor must also know the same information for several reasons. At times there will be no agents involved in a sale and someone else must be doing the activities mentioned here. In another scenario, the investor should know what the licensee is doing to ensure the marketing efforts are in line with the principal's (either the buyer or seller) expectations in finding the appropriate property or qualified purchaser. In either case, the marketing aspects are as important to an owner (or potential owner) as they are to a broker/agent.

Assuming that the reader has studied and given some serious thought to the material presented previously in this book, you should already have an idea of which locale will be best for you. It is now time to discuss implementation of this material. Before we begin the important topic of prospecting, we would like to highlight a few thoughts on personal discipline as a part of marketing activities. The fact is that knowing what actions should be taken and actually following through with those actions are two very different things.

Consider the experience of the new graduate with a degree in soil and crop science from the university. The graduate had landed a job with the university's Agricultural Extension Service and his first assignment was to call on the farmers in the area in order to meet them, and at the same time, to explain some new farming techniques recently developed by the university. The first farmer he contacted was on the front porch of his old rundown house, sitting in a rocking chair and smoking a pipe. The young graduate opened the broken gate and observed a few scraggly chickens in the yard. Peering across the weed-infested cropland he observed a few poor cattle in the pasture. After the introductions were over, the new employee began to explain all about the new farming techniques. The farmer listened politely to the whole presentation. He then leaned back in his old rocking chair and blew out a puff of smoke. "Son," he said, "I don't reckon them new ideas are for me, but I appreciate you telling me all about them. You seem a nice enough fellow, so next time you get a chance to come back, we'll go fishing down at the creek." But the young graduate persisted. "Why aren't you interested in these methods that will increase your crop yields by at least fifty percent and allow you to be a better farmer?" the graduate inquired. The old farmer's reply is the crux of this story. "Well, by golly son, right now I ain't farming even half as good as I know how to!"

Much like that old farmer, many brokers and salespeople know that they should be doing more of the less pleasurable tasks that go with the job. This is not to judge readers in terms of what they should or should not be doing, especially if their personal productivity goals, sales volume, and income standards are currently being met in their real estate practices. However, practitioners who are not satisfied that they are accomplishing all they are capable of or desire in those areas should give some thought to these points. Effective farm and land practitioners who are successful, well paid, and satisfied with their business have few if any magical secrets. These fine professionals have attained their position in the field for the most part by doing the things that are necessary to succeed. Although some of those things are not easy or fun, none need be illegal or unethical. Look to the role models in real estate or any other industry. They are the people who accomplish not just the tasks that are exciting and pleasurable, but those that are necessary to attain personal and professional goals.

Please remember, as you review the activities outlined in this chapter, the intent of this book is not to mold the reader into a clone or robot. In fact, each person can carry out the marketing tasks and responsibilities in a manner that is consistent with the following three factors:

1. Individual personality.
2. The people the practitioner deals with.
3. The conditions in which the practitioner operates at the time.

Previously, this chapter focused on the what and why, not the exact how to carry out the marketing tasks and responsibilities. As a practitioner, you should never be afraid of being your own person, or performing tasks in a manner that is best suited for you—especially as long as you work in a way that brings credit to you, your firm, and your profession. After all, there is an old country saying about being a creative risk taker that is shown in Fig. 5-1.

PROSPECTING

There may be as many definitions for words as there are people using those words. To avoid confusion about what we mean by the term "prospecting," we provide this definition: Prospecting is any planned, systematic method of consistently generating positive contacts with people who may need a practitioner's professional help to buy or sell farm and ranch land. There are some key points in this definition that should be noted. The word *any* suggests there are many ways to accomplish a specified task. This book does not pretend to contain all of the various ways to prospect. As long as a method suits the rest of the definition, the method can be supported. The words *planned* and *systematic* convey the notion that a person does not, in the long run, prospect successfully by accident. Effective prospecting requires forethought and persistence in order to achieve the desired payoff. The word *consistently* is a key word for practitioners. Standing in the middle of the busiest road in your entire market area with a sign that says, "Buy/Sell your land with me" may work once. But just because something works once, you should not conclude that it will always be productive. Remember that this chapter is focused on the long haul. *Positive contacts* are vitally important. If your activities put you in contact with people who say, "Never call on me again, ever," then that technique or target audience is not recommended. The idea is that you want to be working with people who *may need* your services. Prospecting ferrets out

Figure 5-1

> The lead mule is the only one
> that gets a change of scenery!

suspects. Qualifying confirms the suspects as prospects, those people who do require your *professional help.* Practitioners want to interact with individuals who can be made aware of what they have to offer. In addition, you must also explain how your assistance can benefit those parties as well as themselves. If the people contacted only want to learn what you know and are unwilling to work with you in a mutually beneficial arrangement, they are marginal prospects at best and you should review your target audience.

Experienced practitioners can confirm that very little business "falls from the sky." Certainly everyone is grateful for the opportunity that appears with limited effort on their part. However, the reality is that such circumstances will not occur on a consistent enough basis to build a career on. This means that you must develop a means by which to get the exposure necessary to meet the people who need your expertise. If you have done your homework on a marketplace as outlined in the previous chapter, you should already have some product knowledge. The task now is to gain access to the right people with whom to share that knowledge. After all, what is it that you have to offer to a potential buyer or seller? In the final analysis, you can offer two things:

1. The knowledge of real estate and its transfer.
2. The energy, effort, and commitment to help carry out the transfer of real estate.

The practitioner does not actually create anything. Frequently, brokers/agents have been accused by unhappy buyers and sellers of raising or lowering values. But only principals make that happen. At best, practitioners are the analysts and interpreters of values as determined by the interactions of buyers and sellers in the market. The practitioner recognizes the need to be aware of happenings in a specialty area and balance this awareness with the development of a means of access to future participants in the marketplace. This is the key to a practitioner's ability to help buyers and sellers carry out their goals in that market.

There are any number of methods of prospecting. Most of the methods are variations on the theme of meeting potential buyers and sellers. Once these contacts are made, the practitioner must be competent enough to provide a service. The technical information provided in this text plus all the resources mentioned here represent overviews or examples of what you need to know. Developing this knowledge takes time, as does creating a positive impression as a knowledgeable practitioner. One of the common mistakes made by newcomers in the farm and land business is to "make the rounds" of potential business contacts and then become disappointed when nothing develops immediately. Without allowing an adequate period of time for the prospecting to work or making any follow-up contacts, some agents "move on" to another marketplace or leave the industry entirely. Such impatient practitioners still do not have a clue as to why the "techniques" being used did not work. Unfortunately, they may continue in their own ineffective way, depriving themselves and consumers of the most efficient use of their time and talent. In addition, these practitioners tarnish the image of the industry.

Practitioners need to remember that the practice of farming and ranching develops (or as some would say, forces) patience. Agricultural people are more aware of seasons and cycles than most other people, due to the nature of the industry. A farmer who is considering the sale of the old homestead may have to "think on it" for a very long time. Experience and numerous years of brokering farm and ranch land lead us to believe that most people of the land take the sale of their property very seriously. Frequently, these people will only sell in the most pressing of circumstances. Likewise, the farmer or rancher considering the purchase of additional land will most often want to move slowly. After all, this investment represents an expensive burden that must somehow pay for itself. The purchase could also represent a threat to other assets that are already owned on a free and clear basis. In short, when you make prospecting plans with a target audience, the best method is to proceed slowly and easily.

THE AWARENESS CYCLE

Several years ago, the concept of the awareness cycle entered into marketing circles. No credit is taken here for its creation, but it is a worthy perspective for your consideration. As applied to real estate, it is a good reminder that any type of prospecting takes time to make an impression in the minds of buyers and sellers. Under this theory, the first stage of the cycle is **awareness.** This simply means the consumer becomes aware that a particular product or service exists. Think about how often someone has been heard enthusing about a fine place to eat, or asking a friend where those great-looking clothes were purchased. In both cases, the answer may have caught the listener unaware that such a restaurant or clothing store existed. The astute real estate licensee or investor should expect that it will take some time for "the word to get out" that the practitioner is actively involved in the marketplace. This step can be hastened by making announcements using mail-outs or an ad in the local newspaper. However, it is important to remember that it is the practitioner's responsibility to ensure that this important first step occurs. And, just because you have attempted to spread the word about what services you are offering, do not be surprised if it takes considerable time for the news to "soak in." Even after a significant amount of time in the field, there may still be some people who are unaware of your involvement, or who still consider you a newcomer.

A visit with an elderly neighbor within a practitioner's market area is worth retelling. As usual, there was a discussion about prices, yields, and the weather. Then the old farmer asked about those "new folks" who bought the old Starr place not very long ago. Those "newcomers" had been owners for eleven years when that comment was made. Of course, to the farmer they were still new neighbors since he had spent his eighty-plus years on the same land. In addition, his family had been working the same property for two generations before him. Using this story to make the point may be overkill, but keep in mind that your reputation may spread far more slowly than you desire.

The second phase of the cycle is one of **recognition.** This aspect primarily acknowledges that it will take a period of time and often repeated contacts with people before they correlate the broker/agent with real estate. The fact is that most consumers need multiple reminders in order to associate the company or person with the proper products or services being offered. Just think about the repetition of advertising on television and radio. Such exposure gets to the point of overuse, but that is not as great a fear for a newcomer who very much wants to gain identity. Some of the prospecting research in residential real estate indicates that a broker/agent may have to call upon homeowners in a particular target area (sometimes known as a farming area) as many as six or more times before the homeowners will recognize the broker/agent as being involved in real estate. Applying these findings to farm and ranch properties solidifies the point of this chapter: *Prospecting takes time to work.*

The next phase under this concept is one of **trial.** At this point, the potential user of the practitioner's services will actually interact with the practitioner and discuss real estate matters. This may start in an indirect manner. A potential buyer or seller may ask in an almost offhand manner about activity or values. Do not take this comment too lightly. It is true the person may be simply "fishing" for information out of curiosity. However, this happens often, and land sales and values are the topic of much "coffee shop" conversation in many areas. On this note, we strongly urge practitioners to be very careful to avoid disclosing information about sales in an unprofessional manner.

In some states, it is possible for such indiscriminate disclosure of private information to be considered a violation of the law of agency. In a few states where exact financial details are a matter of public record, there is less of a problem with such discussions. Still, in most cases the individual buyers or sellers involved in a transaction will not take kindly to a practitioner's "running off at the mouth" about their business. It is important for you to be current on land values in your market area and it is a responsibility that goes along with doing your job competently. There is nothing improper about discussing this topic as part of your attempt to obtain a listing or encourage someone to purchase. But this is different from allowing people to "pump" you for information for no other reason than that they just want to know. And, also keep in mind that the same people who may desire this knowledge would look at the situation differently if it was their business that you were protecting. There have been many business opportunities lost by practitioners who could not respect this point. There are many, many buyers and sellers who expect, or demand, privacy in their dealings. These same people will refuse to do business with practitioners who cannot follow through on this expectation.

Now that some time has been spent suggesting what *not* to do with the trial phase, exactly how should a person act? Although there are no hard and fast rules that can be stated as absolutes, some very definite methods will be suggested. Any time a person desires to discuss real estate with you, give that request the full attention and respect it rightfully deserves. In your own style and personality, send the message that you are ready, willing, and adequately prepared to be of service to either the buyer or seller. By focusing on the needs of the buyer or seller as well as how you can be of service to them, you'll be starting off on the right foot!

In this introductory period when both parties are striving to learn more about each other, it is important that you do not become too pushy. Be enthusiastic and helpful, but not high-pressured. This step is somewhat a qualifying stage for both parties. Concentrate on finding out what the person's expectations are regarding this transaction. And, always keep this most vital question in mind: "Am I comfortable with what this individual wants me to do?" If the answer to this question is yes, then you are on the right track. However, there is a fair chance that the potential buyer or seller will have unrealistic expectations of you. Our educated guess is that every experienced farm and land practitioner and investor has dealt with at least one, and most likely many buyers or sellers who expect more than is reasonable. It is common for the seller, whose property at the present time has a market value of $X, to be pleased to let someone else own the land for one hundred and fifty percent or two hundred percent of the true value. Likewise, the buyer who desires to pay fifty percent or less of value is a known entity. Here is a key point to remember about this type of buyer or seller: They do not represent the type of business opportunities upon which to build a career. In fact, some of these people are best referred to your strongest competitor in the hopes that the competition's time will be wasted by them! Be very cautious when agreeing to work with people who expect unrealistic performance from you. If you do not feel that you can reasonably carry out the buyer's or seller's expectations, acknowledge that, explain why the opportunity is being declined, thank them, and then offer your willingness to be available in the future. One might think that such an approach will result in a lost business opportunity. In fact, often this is simply not true. A buyer or seller who has unrealistic expectations is not a good opportunity. If you can make a good case for declining to help the buyer or seller under the present conditions, you may indeed be helping to ensure future business.

There are only a few possible outcomes of such a scenario. The person may find someone able to perform (versus try and fail). In that case, you will have learned a valuable lesson in terms of what is possible. More likely than not, if you know your marketplace as well as you should, what will transpire is that no other agent will be able to meet the person's expectations either. When that is the outcome, as it frequently is, who are the heretofore unrealistic buyer or seller going to make contact with again? Experience shows that it will be the true professional who told them not what they wanted to hear, but rather what they needed to hear. This becomes a part of their education on local and current real estate values. Remember, the focus of this book is on providing long-term service in a particular marketplace. The approach outlined here when someone "tests" the practitioner is based on many years of experience dealing with all sorts of buyers and sellers of farm and ranch land. The key point is for you to know your abilities as well as your limitations. When you commit to perform a real estate service, do so with as much knowledge as possible of your marketplace. Work as hard as possible to serve that individual. Do not hesitate to admit an unwillingness to work with someone who expects more than is realistic. And seek those buyers and sellers who are willing to work with you within realistic expectations.

The next phase of the awareness cycle is **preference.** In this case a buyer or seller expresses a preference to do business with a certain practitioner. Just because someone has decided to use your services, there is no reason to become cocky and overconfident. You will still need to be able to perform at your highest level possible. When working with a seller in terms of listing a property and trying to sell it, you need to obtain all the information referenced in the property checklist. You must gather as much current and accurate detail as possible for each listing. A common mistake is for a listing agent to rush to begin marketing a new property. The best rule of thumb is to proceed slowly to promote the property and to dig deeper during the information gathering period.

It is very difficult to imagine a serious, qualified buyer complaining that the listing agent provided too much relevant information. Be forewarned that a poorly prepared listing agent will receive plenty of heat. Some of this heat may be in the form of justified criticism; however, it could also be in the form of a "failure to disclose" lawsuit. Each and every state has some form of Deceptive Trade Practices Act and you should become familiar with the laws in your state. It appears that many of the problems encountered by practitioners in this area center on the fact that they are inadequately informed and unprepared to convey enough relevant information to a consumer. (See Fig. 5-2.)

The same concept applies if a practitioner is working with a buyer and representing their particular interests. You must obtain from listing agents, sellers, or through your own research as much accurate detail as possible to assist the buyer in making the most prudent decision possible. Do not hesitate to acknowledge, at least to yourself, a lack of understanding regarding a particular situation. A recent example illustrates this point. A tract of land listed for sale was situated in the middle of a rapidly escalating mineral development area. The cause for such strong interest was based upon new technology that allowed for the recapture of certain minerals that were previously inaccessible. It goes without saying that such technology is very sophisticated, as well as expensive. There is very little chance that a layperson with limited experience in the drilling industry would be able to adequately explain the monetary implications or impact on the land of such a procedure. Rather than pretend, it is prudent to admit the shortcoming to potential buyers and have an explanation drawn up by a competent professional from that field in order to help the prospective buyer properly evaluate the land. Through the use of diligent and thorough information gathering, a real estate professional earns and keeps the respect and confidence of buyers and sellers.

The final stage of the awareness cycle is **insistence.** In this case, the practitioner has so well served a person in the past that there is no doubt as to who they will turn to with additional real estate needs. What a wonderful position to be in, when your ability is so well appreciated and rewarded by such loyalty! While each of these relationships is unique enough to be called special, they are

Figure 5-2

REALTOR®

SAN ANTONIO BOARD OF REALTORS, INC.

9110 IH 10 WEST, SAN ANTONIO, TX 78230

SELLER'S DISCLOSURE NOTICE

EQUAL HOUSING OPPORTUNITY

NOTE: Effective January 1, 1994, Section 5.008 of the Texas Property Code requires a seller of residential real property of not more than one dwelling unit to deliver a copy of the Seller's Disclosure Notice, completed to the best of the seller's belief and knowledge, to a purchaser on or before the effective date of a contract for the sale of the property. If a contract is entered into without the seller providing the notice, the buyer may terminate the contract for any reason within seven (7) days after receiving the notice. If information required by the notice is unknown to the seller, the seller may indicate that fact on the notice and thereby comply with the requirements of Section 5.0008 of the Texas Property Code.

CONCERNING THE PROPERTY AT ______________________________

(Street Address and City)

THIS NOTICE IS A DISCLOSURE OF SELLER'S KNOWLEDGE OF THE CONDITION OF THE PROPERTY AS OF THE DATE SIGNED BY SELLER AND IS NOT A SUBSTITUTE FOR ANY INSPECTIONS OR WARRANTIES THE PURCHASER MAY WISH TO OBTAIN. IT IS NOT A WARRANTY OF ANY KIND BY SELLER, SELLER'S AGENTS, OR ANY OTHER AGENT.

Seller ☐ is ☐ is not occupying the Property. If unoccupied (by Seller), how long since Seller has occupied the Property? __________

1. The Property has the items below: [Circle Yes (Y), No (N), or Unknown (U), even if you intend to exclude the item from the sale]

Y N U	Range	Y N U	Oven	Y N U	Microwave
Y N U	Dishwasher	Y N U	Trash Compactor	Y N U	Disposal
Y N U	Washer/Dryer Hookups	Y N U	Window Screens	Y N U	Rain Gutters
Y N U	Security System	Y N U	Fire Detection Equipment	Y N U	Intercom System
Y N U	TV Antenna	Y N U	Cable TV Wiring	Y N U	Satellite Dish
Y N U	Ceiling Fan(s)	Y N U	Attic Fan(s)	Y N U	Exhaust Fan(s)
Y N U	Central A/C	Y N U	Central Heating	Y N U	Wall/Window A/C
Y N U	Plumbing System	Y N U	Septic System	Y N U	Public Sewer System
Y N U	Patio/Decking	Y N U	Outdoor Grill	Y N U	Fences
Y N U	Pool	Y N U	Sauna	Y N U	Spa or Hot Tub
Y N U	Pool Equipment	Y N U	Pool Heater	Y N U	Auto. Lawn Sprinkler System
Y N U	Fireplace(s) & Chimney (Woodburning)			Y N U	Fireplace(s) & Chimney (Mock)
Y N U	Gas Lines (Nat./LP)			Y N U	Gas Fixtures

Garage:	Y N U	Attached	Y N U	Not Attached	Y N U	Carport				
Garage Door Opener(s)	Y N U	(Number)	Y N U	Control(s) Number						
Water Heater:	Y N U	Gas	Y N U	Electric	Y N U	Solar	Y N U	Other		
Water Supply:	Y N U	City	Y N U	Well (Private)	Y N U	MUD	Y N U	Co-op	Y N U	Other
Air Conditioning:	Y N U	Zoned	Y N U	Electric	Y N U	Gas				
Heating:	Y N U	Zoned	Y N U	Electric	Y N U	Gas	Y N U	Solar		
Security System:	Y N U	Owned	Y N U	Leased	If leased, lessor's name: ______					
Satellite Dish:	Y N U	Owned	Y N U	Leased	If leased, lessor's name: ______					
Water Softener:	Y N U	Owned	Y N U	Leased	If leased, lessor's name: ______					
Roof:	Type: ______				Age: ______ (approx)					

Are you (Seller) aware of any of the above items that are not in working condition, that have known defects, or that are in need of repair? ☐ Yes (if you are aware) ☐ No (if you are not aware). If yes, then describe. (Attach additional sheets if necessary): ______________________________

____

____

____

2. Are you (Seller) aware of any known defects/malfunctions in any of the following? Circle Yes (Y) if you are aware, circle No (N) if you are not aware.

Y N	Interior Walls	Y N	Ceilings	Y N	Floors
Y N	Exterior Walls	Y N	Doors	Y N	Windows
Y N	Roof	Y N	Foundation / Slab(s)	Y N	Basement
Y N	Walls/Fences	Y N	Driveways	Y N	Sidewalks
Y N	Plumbing/Sewers/Septics	Y N	Electrical Systems	Y N	Lighting Fixtures
Y N	Other Structural Components (Describe): ______				

____

If the answer to any of the above is yes, explain. (Attach additional sheets if necessary): ______________________________

____

SABOR FORM 545 (TAR-044) Rev. 10/12/94 SELLER'S DISCLOSURE NOTICE by no means rare. From this kind of support, a salesperson or broker can derive repeat as well as referral business. In dealings with this most important group of people, avoid the temptation to try to be all things in real estate for them. What we mean by this is that there may be times when these people need real estate services that you are not qualified or experienced enough to perform. You will do yourself a great benefit by introducing these people to a specialist with greater skills who can be of better service to them. We strongly recommend that you put the buyers' or sellers' needs ahead of yours and carefully explain your logic for this action. You may find you have a stronger rapport and better business tie with these people after such a move.

As you continue to study and practice the marketing activities associated with farm and ranch real estate, it is important to keep in mind the concept of the awareness cycle. It is important for you to know *where* you stand with each business opportunity and *how* to progress to the next step, concluding in a satisfactory property exchange to the benefit and satisfaction of all the parties involved.

AGENCY ISSUES

We made a conscious choice in writing this text to include only brief information regarding selling skills. Topics like handling objections, buying signals, closing techniques, and negotiation strategies are covered extensively in a variety of real estate sales and marketing texts. But one issue very near and dear to the heart of anyone involved in marketing bears mentioning. In this section we address issues regarding agency. Please keep the following in mind as you read this portion of the book:

1. This topic can change rapidly based upon new court decisions, so it is imperative that you keep current.
2. What may be correct and legal in one state could easily be different in another regarding the format and nature of proper disclosure.
3. The authors are not attorneys and are making no claim to offer legal advice on any aspect of real estate.

Even in this fairly safe setting of a book, we felt such caveats should be included. But the potential for litigation is a reality in real estate, maybe more today than ever before. No licensees go into a real estate opportunity with the hope of having a lawsuit filed against them. But as dear grandmother was fond of saying: "An ounce of prevention beats a pound of cure any day." The opinion expressed in this text is that some and maybe even most of the pitfalls regarding disclosure can be avoided by following the fundamental guidelines presented here.

Agency

A major disclosure area is that of proper explanation and clarification of who the licensee represents. Given that each state will vary on this issue, we offer a somewhat basic and generic review. A customer is someone a licensee will work **with** while a client is someone to work **for.** There is a great deal of confusion about this important difference in the minds of many consumers. For example, just because a licensee helps a buyer/customer to find a suitable property, prepares and presents an offer, assists in securing financing, and handles the details of consummating the transaction, this does not mean that the practitioner was the buyer's agent. Remember, the licensee owes the buyer honesty and cannot misrepresent property under the guise of being a seller's agent.

On the other hand, as an agent, the licensee owes the principal/client much more. There are many ways to outline the duties of an agent to a principal, but this acronym may be helpful as a reminder:

Loyalty—remember whose interests the licensee is there to protect.

Accounting—keeping accurate track of funds entrusted to the licensee.

Notice—make certain the principal is fully informed.

Disclosure—tell all parties who you are working for and let your actions show it.

Confidentiality—keep the principal's views to yourself.

Obedience—follow all of the principal's legal requests.

Performance—do your job to the best of your ability.

LANDCOP carries a heavy weight of responsibility. No law makes the licensee work for anybody. If the principal's requests are not in line with your company's policies, then decline the opportunity to work for this person. But once you begin working for a principal, LANDCOP is the standard.

Types of Relationships

There are a variety of agency combinations possible in a real estate transaction. No matter which relationship the licensee enters into, experience indicates it is very important to have whatever agreement is reached with a consumer committed to writing. In some instances, state laws will not allow a practitioner to bring suit for a commission unless there is a *written* agreement between the agent and the principal. This authorization needs to clearly outline the specific terms and conditions of the duties and obligations of all parties.

Traditionally, the most frequent form of representation is that of an agent representing a seller. A listing agreement is commonly used as the written document that formalizes this agreement. In this case, the buyer would be a customer when dealing with the listing agent. Keep in mind that in many states the "true agent" is the broker of the firm and not the licensed salesperson employed by the broker. Licensees should make certain they are aware of the various types of listings that are allowed in their state. Exclusive agency, exclusive right to sell, nonexclusive (sometimes referred to as open), net, and multiple are the common types of listing agreements that most practitioners are accustomed to using. Each type of listing agreement also has its differences and variations. You should make certain you know which type of listing agreement your office policies allow you to use.

Another form of seller representation is subagency. In this case, a licensee from another firm (no subagency exists within the same firm in any state that we are aware of) agrees to work on behalf of the seller *through* the listing broker. This practice was widened considerably when multiple listing services became popular. Some states hold the seller and/or listing agent vicariously liable for actions of the subagent. This means that licensees are responsible for what the subagent says and does when presenting their property or listing. Certain states have modified this to say that only when a seller or listing agent know of the subagent's error and do nothing about it is there liability. This "extended responsibility" has caused some sellers and real estate companies to stop offering subagency.

The buyer (or tenant) representation relationship has been around a fairly long time in commercial real estate, but is newer in other specialty fields. In this case, the agent serves the buyer as a client and looks after his or her best interests while treating the seller as a customer. In rural land this type of relationship has been seen when an agent represents a large company with farmland in many areas. It is even more common when an agent seeks land for and represents a developer. In a case like this there frequently are a buyer, buyer's representative, seller, and seller's representative involved. It will not be long before a program is required to know the players!

An additional example of an agency relationship occurs in what has been called single agency. This occurs when a licensee represents a particular consumer no matter whether that person is a buyer or a seller. Such circumstances happen most often when there is a long-term orientation of providing whatever service is needed for the particular individual or company. In this case, licensees are cautioned not to endeavor to be more than their abilities warrant. For example, we know of a broker who does some work with a farming organization that specializes in one crop. This fine firm has holdings in several states ranging from North Dakota to Florida and from Missouri to Texas. While serving the needs of this firm is pleasurable and profitable, it would be unwise to think one broker automatically could handle all their work in all those states. One person could not handle even all their work in one state if their commercial ownership and leases over and above their agricultural holdings are considered. The key in this form of agency representation is to know your limits. Again, it is important to remember that "pigs get fat, but hogs get slaughtered." Licensees are reminded not to overstep their abilities.

The final form of agency relationship addressed here is the "hybrid form." This may be called dual or intermediary agency. It may even be known as facilitator or middleman in some states. Remember, the terms may vary. Whatever it is titled, the concept is to represent **both** parties in the transaction or to represent **neither** party. This is a particularly difficult area to discuss on a broad scale. Some states do not recognize multiple representation in the same transaction. On the other hand, in some parts of the country it has been decided that an agent must represent **somebody** (buyer or seller, tenant or landlord) in each transaction. Practitioners should be absolutely certain they know what is allowed in their state.

It seems obvious for the licensee to clarify what form of agency is to be practiced and who the agent is going to represent. Some states now require that this disclosure be made, either verbally or in writing, at the first face-to-face meeting. Other states require the disclosure prior to a purchase money contract being finalized. Our thought is that very little in real estate should be left to just a verbal discussion. We further believe that information should be offered earlier rather than later. Therefore, the recommendation of this book is to clarify any and all agency issues very early and always in writing.

As a helpful example, a copy of the agency disclosure document now required in Texas is presented in Fig. 5-3. Practitioners should make certain that they are aware of and are consistently following the laws of their particular state as well as the practices and polices of their office. This aspect of real estate is simply too important to be taken lightly or ignored.

Figure 5-3

Approved by the Texas Real Estate Commission for Voluntary Use

Texas law requires all real estate licensees to give the following information about brokerage services to prospective buyers, tenants, sellers and landlords.

Information About Brokerage Services

Before working with a real estate broker, you should know that the duties of a broker depend on whom the broker represents. If you are a prospective seller or landlord (owner) or a prospective buyer or tenant (buyer), you should know that the broker who lists the property for sale or lease is the owner's agent. A broker who acts as a subagent represents the owner in cooperation with the listing broker. A broker who acts as a buyer's agent represents the buyer. A broker may act as an intermediary between the parties if the parties consent in writing. A broker can assist you in locating a property, preparing a contract or lease, or obtaining financing without representing you. A broker is obligated by law to treat you honestly.

IF THE BROKER REPRESENTS THE OWNER:
The broker becomes the owner's agent by entering into an agreement with the owner, usually through a written listing agreement, or by agreeing to act as a subagent by accepting an offer of subagency from the listing broker. A subagent may work in a different real estate office. A listing broker or subagent can assist the buyer but does not represent the buyer and must place the interests of the owner first. The buyer should not tell the owner's agent anything the buyer would not want the owner to know because an owner's agent must disclose to the owner any material information known to the agent.

IF THE BROKER REPRESENTS THE BUYER:
The broker becomes the buyer's agent by entering into an agreement to represent the buyer, usually through a written buyer representation agreement. A buyer's agent can assist the owner but does not represent the owner and must place the interests of the buyer first. The owner should not tell a buyer's agent anything the owner would not want the buyer to know because a buyer's agent must disclose to the buyer any material information know to the agent.

IF THE BROKER ACTS AS AN INTERMEDIARY:
A broker may act as an intermediary between the parties if the broker complies with The Texas Real Estate License Act. The broker must obtain the written consent of each party to the transaction to act as an intermediary. The written consent must state who will pay the broker and, in conspicuous bold or underlined print, set forth the broker's obligations as an intermediary. The broker is required to treat each party honestly and fairly and to comply with The Texas Real Estate License Act. A broker who acts as an intermediary in a transaction:

(1) shall treat all parties honestly;
(2) may not disclose that the owner will accept a price less than the asking price unless authorized in writing to do so by the owner;
(3) may not disclose that the buyer will pay a price greater than the price submitted in a written offer unless authorized in writing to do so by the buyer; and
(4) may not disclose any confidential information or any information that a party specifically instructs the broker in writing not to disclose unless authorized in writing to disclose the information or required to do so by The Texas Real Estate License Act or a court order or if the information materially relates to the condition of the property.

With the parties' consent, a broker acting as an intermediary between the parties may appoint a person who is licensed under The Texas Real Estate License Act and associated with the broker to communicate with and carry out instructions of one party and another person who is licensed under that Act and associated with the broker to communicate with and carry out instructions of the other party.

If you choose to have a broker represent you,
you should enter into a written agreement with the broker that clearly establishes the broker's obligations and your obligations. The agreement should state how and by whom the broker will be paid. You have the right to choose the type of representation, if any, you wish to receive. Your payment of a fee to a broker does not necessarily establish that the broker represents you. If you have any questions regarding the duties and responsibilities of the broker, you should resolve those questions before proceeding.

Real estate licensee asks that you acknowledge receipt of this information about brokerage services for the licensee's records.

Buyer, Seller, Landlord or Tenant　　　　　　Date

Texas Real Estate Brokers and Salesmen are licensed and regulated by the Texas Real Estate Commission (TREC). If you have a question or complaint regarding a real estate licensee, you should contact TREC at P.O. Box 12188, Austin, Texas 78711-2188 or 512-465-3960.

EQUAL HOUSING OPPORTUNITY

6 Listing, Promoting, and Showing Property

If you keep doing the right things (versus worrying about doing things right) in terms of trying to develop business opportunities, some business will come your way. This chapter will concentrate on several important aspects of marketing. One critical aspect is to do a good job when listing property. Once the property is listed, you must promote it in order to attract potential buyers. Of course, little good can come from the first two steps if you are incapable of presenting the property in a competent manner. All of these activities go hand in hand in successful marketing. It should be emphasized that there are some very fine real estate books that address each of these topics separately in greater depth than is provided in this book. Use those other resources for foundational support. The primary focus here is on farm and ranch land.

THE IMPORTANCE OF LISTINGS

Most knowledgeable real estate practitioners will agree that the company (or person) in control of the listings in a trade area will be the one who is in control of the marketplace. Either through their own initiative or by being brought to the practitioner through other licensees, buyers will present themselves to a broker/agent who has well-priced, attractively promoted listings. There is little doubt that the axiom shown in Fig. 6-1 holds true in real estate.

If you agree with the logic of this viewpoint, you should also agree to commit yourself to developing the skills necessary to become competent at listing properties for sale.

In order to ensure that everyone is "on the same hymn page" regarding this important marketing activity, a listing is defined as: A legally binding contract that allows a licensee to attempt to market property owned by the seller. There are many different types of listings to be found throughout the various states. It is beyond the scope of this book to present an in-depth analysis of the various forms or types of listings utilized in different areas of the country. Some of the more common types of listings and a brief explanation of each follow:

1. Nonexclusive Listing (sometimes known as an open listing): The seller may authorize any number of licensees to attempt to market the property and the seller is only obligated to pay the licensee who obtains a ready, willing, and able buyer. Further, the seller may sell the property directly and owe no brokerage fee to anyone.
2. Exclusive Agency Listing: The seller selects only one licensee to market the property and all other licensees must work through the listing agent. However, the seller reserves the right to sell the property directly and owe no brokerage fee.

Figure 6-1

> If your desire is to last in this business, you'd better learn how to properly list.

3. Exclusive Right to Sell Listing: There is only one broker and the same rules apply for cobrokering as are explained above. The seller relinquishes the right to sell the property alone, and owes a fee to the listing agent no matter who sells the property.
4. Multiple Listing: A group of licensees band together to form a joint marketing effort with a guarantee of cooperation among the membership. Usually, this joint marketing effort is coordinated by the local Board of REALTORS.
5. Net Listing: List price is based upon a set amount due to the seller with all other associated costs and fees being added to that amount.

Some of these types of listings are not applicable in all areas and there may be other types being used. Practitioners are strongly encouraged to become familiar with the local practice and laws applicable in their state. Furthermore, practitioners are urged to become competent in the use of the proper documents necessary to create a valid listing agreement. The listing agreements or contracts vary widely from state to state and practitioners must learn the forms that are used and recognized in their area.

THE LISTING PRESENTATION

Actually being called upon by a seller to make that first listing presentation is an experience few practitioners forget for a very long time. It may help ease some of the butterflies to remember that while each person has a different style and approach, most listing presentations will consistently have the same main parts. These are discussed in this chapter.

SELLER EXPECTATIONS

Basically, most sellers desire the same three things when selling their property. The seller wants:

1. The best possible price.
2. In the fastest possible time.
3. With the least amount of inconvenience to the seller possible.

These three goals should not come as a big surprise to anyone. After all, nearly everyone would desire these same three things if they were selling. The astute practitioner will focus on those three factors when making a listing presentation. A common mistake that newcomers make is to overcomplicate the presentation. If the practitioner concentrates on these few things there is less chance that something will go wrong. It is important to explore these goals one at a time. As you read the next portion of this chapter, keep in mind that there is a detailed explanation of the appraisal process in another chapter of this text.

The Best Possible Price

The task of a real estate practitioner is not one of setting values for property. Rather, the practitioner's task is one of researching and reporting values to potential buyers and sellers. A common method for accomplishing this is the use of a Competitive Market Analysis, or CMA. This tool may go by other names in various parts of the country but whatever it's called, its purpose is to provide a history of the market area. This history breaks down what is for sale, what has sold recently, and what did not sell.

The **what is for sale** portion represents alternatives for purchasers and competition for sellers. This reflects what sellers want or hope to receive, and while it is valuable to know it is only a starting point in the market analysis. The **what has sold recently** portion is very important in that this portion of the history represents what sales actually occurred. Rather than focusing on what sellers desired or buyers preferred to pay, this section addresses actual sale prices and is often a better indication of value than the other two sections.

The idea of what is recent for farm and ranch land may differ from that in other areas of real estate. The time frame will vary depending on the local area and its activity. However, in most parts of the nation, a sale that is one or two years old may be worthy of note, assuming no drastic change has come about that would make such a sale invalid. It should be emphasized that a competent practitioner will be as certain as possible to quote accurate selling price figures in all situations. But in times of major change in the local marketplace accuracy becomes even more critical. The last part of a CMA is the **what did not sell** section, which lists properties that were for sale but for whatever reason did not sell and were withdrawn from the marketplace. Do not assume that the only reason such properties did not sell was improper pricing. Other contributing factors could include poor marketing efforts, an uncooperative seller, poor timing of sales efforts, and general changes in the area's economy.

No real estate practitioner can consistently predict, much less dictate, what properties will sell for with complete accuracy. But armed with a CMA, a generally accurate opinion on value range can be determined. The more experience and success in the business you attain, the more comfortable you will become with the next statement. All sellers expect the best possible price, but if their expectations cannot be met within the indicated value range, you may not want their listing. The idea of turning down a listing because it violates the practitioner's acceptable range may seem strange, but remember that in this scenario, your price opinion was based upon a complete, current, and accurate CMA. For how much above the indicated value should a practitioner agree to take a listing? Many farm and land brokers and salespeople have different views on this point. Some believe that unless the property can be listed for no more than fifteen percent above the indicated value, the listing is not worth spending time on. Others will not take a listing at any price over their recommended amount.

The Fastest Possible Time

One of the reasons for the above statement is that almost everyone wants to **sell** real estate, not just **list** it. There is little doubt that properly priced listings will generally sell faster. Please understand that we do not advocate listing for an inappropriately low amount just to obtain a quick sale. That is far from the main point we are trying to convey here. Having a property on the market involves many aspects, some good and some not so good. One of the not so good aspects for many sellers is that having their place up for sale is, frankly, a pain in the neck. If a seller lives on the land, there is the feeling of "living in a fish bowl." And while keeping a home clean and presentable in the city is a challenge, it is definitely more difficult in the country! Even if that is not a problem, most people of the soil just do not like to have anyone on their property they have not specifically invited.

This somewhat uncontrolled access "sticks in the craw" of sellers time and time again. The astute practitioner recognizes this fact and emphasizes the merit of properly pricing the property in order to expedite a sale.

MARKETING/CLOSING COSTS AND NET PROCEEDS TO SELLER

This may come as a shocking statement to the uninformed, but virtually all experienced practitioners who are successful in listing property will agree with the following sentence: For most sellers the real issue is not what the property will sell for from a gross sales price perspective, but rather the amount they will receive from the sale. Certainly effective pricing is critical. But many other factors also play major roles in successfully marketing or investing in agricultural property. For example, if the seller's only focus is sales price, then consider whether a practitioner should seek an offer for the land based upon that singular factor if the following terms and conditions apply:

1. The seller is offered exactly the desired price for the property.
2. No down payment will be made on the purchase and the seller will pay all closing costs associated with title transfer.
3. Furthermore, no payments will be made for the first ten years of ownership and only after ten years will the payments begin, with the note being amortized over a fifty-year period and carrying a simple interest rate of two percent.
4. At any time the property may be sold and the note will be fully assumable by any buyer.
5. Should the current or any future buyer desire to pay off the existing note, they can do so at any time without advance notice.
6. Finally, the sellers agree to subordinate their note to any other note that may be placed upon the land in the future.

Now the previous example may be overkill, but consider the key point and learn from it. The sale of a piece of land has any number of costs associated with it and some of those costs can be fairly expensive under certain conditions. Furthermore, the seller must consider certain income and tax factors when evaluating any offer. The above-referenced offer is so far out of line as to be beyond insulting and approaching hysterically funny. A licensee or investor buying or selling real estate should have a thorough familiarity with the closing costs involved with title transfer and be able to communicate that information to others.

One trick of the trade that has served well over the years is the development of a small glossary of terms for buyers and sellers that provides an outline of what common closing costs they can expect. Practitioners should be certain to use current terms and dollar amounts or estimates as relevant to their respective areas. It is also important to emphasize to the buyers and sellers that the closing cost information provided at the outset of a business transaction is strictly an estimate at the present time. Explain that specific dollar amounts can only be arrived at when all details are solidified and the transaction is very close to being finalized with the actual title transfer. Practitioners need to be careful not to overstep their role in terms of acting as a substitute accountant or attorney. If there are technical questions of a tax or legal nature, you should admit to your limitations and encourage the impacted party to seek other professional help. A professional in the marketing of farm and ranch land can do many important things to facilitate buying and selling of property; however, there are some things that require the expertise of professionals in other fields.

AGENT CONTRIBUTIONS

Consider the following scene: You are sitting in the kitchen of a landowner who is seriously evaluating several real estate companies before selecting one firm to market her very desirable farm. After coffee and small talk, the seller looks the practitioner straight in the eye, and says, "Enough chitchat, let's get down to business. Exactly what is it you can do for me to help sell this place?" What a wonderful opportunity, but also a tremendous responsibility! Every practitioner hopes to be in this exact situation. However, this situation causes a little uneasiness in even the most experienced practitioners. Inherent in any well-compiled listing presentation should be a presentation of what you can offer the seller. People in various areas will respond differently based upon local laws, practices, office policies, and individual preference. But each and every person involved in marketing real estate will be asked to answer this question sooner or later. You had better be prepared to do a good job of responding to this question.

The example provided here is one that represents a starting point for each practitioner's own adaptation. This example can be customized by expanding or altering it to fit each individual's needs. It is important to consider the merit of putting whatever is finally chosen into a handout format much like the sample depicted in Fig. 6-2. Creation of such a form for the seller serves two purposes:

1. The form can serve as a nice prompter or script whenever the practitioner is asked the question, "What can you do for me?"
2. The form also serves as a written document that can be left with the seller as a reminder and reinforcement of the practitioner's pledged contributions.

Remember, when sellers ask that question, they are exhibiting a normal human response. Practitioners should address the question as clearly and confidently as possible in their presentation.

PROMOTING PROPERTY

Congratulations! You now have a listing of your own. Now what? One veteran farm and ranch broker with many successes and a few failures regarding her listings over the years stated it very well. She said, "When you buy a dog, you also get its fleas!" The chance to help a seller merchandise a property has finally become the practitioner's responsibility. Such a responsibility also carries several heavy-duty obligations on the practitioner's part. The purpose of this material is to review some of the overall purposes that promotion serves when marketing farm and ranch land. This chapter will offer practitioners specific examples to help them think creatively

Figure 6-2

The G. MACK Company REALTORS Marketing Plan

The following are the details we will handle for you in order to facilitate the sale of your farm or ranch:

1. We will use current and accurate local land sales to help you determine the correct listing price for your property.
2. Our extensive advertising program will ensure that your property receives proper exposure.
3. Our referral and multiple listing services will involve other farm and land brokers in the marketing efforts.
4. Our **well-trained** salespeople are familiar with **local** properties and are available to handle all inquiries.
5. We will qualify interested parties **before** showing your property.
6. We will represent **you** through all negotiations until the transaction has been completed.
7. We will monitor all the details of the transaction from the beginning to its final conclusion.
8. We will follow up and follow through even **after** the sale.

WE SELL BY DESIGN—NOT BY ACCIDENT
YOU CAN COUNT ON US

about how to promote their inventory in a more effective manner, whether as a licensee or as a seller. Once potential buyers are attracted to the listing, how can the property be most effectively presented? Presentation is another challenge and a very important aspect of marketing. The following material will provide you with various showing guidelines that are worthy of consideration.

A Brief Overview of Promotion

There are numerous fine books written on the subject of real estate promotion and its various dimensions. The intent here is to review some of the most basic of the objectives without duplicating other resources. Some, but not all, of the objectives of promotion will be addressed here. The objectives included are introducing or continuing the exposure of a company, promoting a specific property, and attracting buyers or sellers. Each of these objectives is addressed below.

Introducing the Company

When a new company is created, an announcement of this event needs to be made. Even when the people involved are well known in the area, the firm will still need to be formally introduced to the area. A new company formation is illustrated in Fig. 6-3.

Practitioners are advised to mention each of the people and their respective roles within the new company in whatever promotional campaign is utilized. Also, it is important for the principals to state where this new company is physically located. A specific invitation to stop by to see the new location is also a very good idea. The intent of this kind of advertisement is to put forth an image of new energy and ability by highlighting the people involved.

Figure 6-3

WE ARE HERE FOR YOU

Effective June 1, Jodaro Realty Enterprises announces its

GRAND OPENING!

The owners, veteran real estate brokers Caleb Elks, Ethan Elks, and Seth Davis, are ready to serve your farm and ranch needs. With over twenty-five years of combined real estate marketing experience, Caleb, Ethan, and Seth have much to offer you.

Jodaro Enterprises' staff also include Dena Rae, Secretary and Lovell Harrington, Office Manager. You are invited to stop by for a visit. The office is located at 31 RED ROCK RANCH LANE and the office telephone number is 512/334-6254.

Continued Exposure of an Existing Firm

Contrast the illustration in Fig. 6-3 with the one in Fig. 6-4. In the case of an older, established firm, the emphasis is very different. The intent here is simply to remind the target audience that the practitioners have been around and will be around for a long time. It is almost like making the statement, "We have a grand announcement to make" and then announce there is nothing new with the company. For many people, such stability is comfortable and dependable, so why not emphasize the positives that such a long-running history of service can entail? Agents who find themselves involved in a mature organization should not become lazy! Please note that the objective is to provide **continued** exposure.

Promoting a Specific Property

Many different methods exist for calling attention to a specific property. With farm and ranch land the more common methods include classified advertising and developing a property information

Figure 6-4

THREE GENERATIONS STRONG

Black Kettle Realty has served the Cheyenne/Sayre area since 1954. Three generations of our family have lived and worked in this community for more than sixty years.

WE KNOW YOU AND THE MARKETPLACE

Our ONLY real estate focus is property located in and around the Cheyenne/Sayre area. We will be in business through the good and bad times as well as the ups and downs of our local economy.

When you need real estate services, remember that we have been and will be here for you. Call us anytime if we can help you. K. B., Kendra, and Ariel Mackey. Our telephone number is 405/733-2852.

packet. In order to help the agent understand some key ideas related to this important marketing task, we will use a specific piece of property that was an actual listing as an illustration.

When promoting a particular property through the classified advertising channel, the first thing that practitioners should determine is the **target audience** they wish to reach. Figure 6-5 accomplishes this immediately in the heading of the advertisement.

Is there any doubt who the advertisement is aimed at? The buyer who is looking for irrigated farmland may not so much as read the advertisement, but that does not matter because the people who the practitioner wants to notice and read the advertisement are quickly (maybe blatantly) identified within the advertisement itself. It may be that the same property would be attractive for different reasons to another target audience wanting something vastly different, but that would be the exception rather than the rule for most properties. Remember the appraisal principle of Highest and Best Use. The real estate marketer must have a similar concept in terms of identifying **who** will be most likely to want this particular property.

The next challenge that the practitioner must address is **where** this target audience is located. In keeping with the same ranch example, where should it be promoted? It is unlikely that a local person, who is attempting to make a living by raising crops or

Figure 6-5

HEY, TROPHY HUNTERS!

1,285 acres of prime hunting country. All heavy brush except for 25 acres of food plots. Three stock tanks. Good road system. Mineral control. For more information call today.

NOVER/SALYERS REALTY

Joseph or Greg Salyers—602/466-3811
Pat, Trish, or Helen Nover—602/545-4115

YOUR HUNTING COUNTRY SPECIALISTS

domesticated animals, would be a prime candidate. If the cost is not prohibitive, it is wise to place an advertisement in the local paper, just to have all of the bases covered. Most likely, however, the better use of promotional money would be in a specialized magazine devoted specifically to the topic of game management or trophy hunting. Unfortunately, such an advertisement may prove to be too costly.

Based upon experience, the target audience will be found in or close to a major metropolitan area, which is where practitioners are counseled to spend the majority of their money for this particular property if at all possible. Unless you are associated with a veteran broker in the field, as a newcomer to farm and ranch real estate, the determination of where best to place advertisements will have to be a trial and error process. While experience is the best teacher, the College of Hard Knocks has a very high tuition payment.

There is clearly some merit in seeking help from competent sources. This could include observing where other practitioners place their advertisements, asking the associates within your own firm, or determining from other buyers or sellers what they read. Above all, practitioners must not forget to monitor the response rate generated by their advertisements. Remember, promotional effectiveness is not based upon the practitioner's personal likes and dislikes but upon buyer or seller response. As one experienced REALTOR stated, "A good advertisement is easy to recognize. It is the one that makes the phone ring."

Before we move on, it is important to review a few more basics for writing advertisements in order to help you to get the most out of your advertising expenditure. You are advised to avoid providing too much information in your advertisements. People frequently read advertisements to eliminate properties that do not meet their needs. Every additional piece of information within the advertisement increases the risk of a particular property not meeting a potential buyer's criteria. It is also advisable not to use terms that may not be understood by potential prospects. This includes words common only to a certain area or slang or an excessive use of abbreviations. Figure 6-6 is an example of all these elements combined into a single advertisement.

Figure 6-6

WHAT A DEAL!
452 acres, w/372 in crp., 1/2 min. & 1/2 exec. rts. Good roads, mill, pens, tank. Has birds. O/C @ 10 for 10 w/30% down.
BLACK GOLD REALTY—817/624-5875
Ask for Eric or Robert Hoffert

Is it possible to figure out what the creators of such an advertisement were trying to accomplish? Perhaps they wanted to tell potential buyers that three hundred and seventy-two acres of this property was in a Crop Reduction Program (a federal set-aside program) which generates some stable income for the property. Furthermore, they wanted prospective buyers to know that a buyer would receive fifty percent of the mineral leasing authority, as well as fifty percent of the minerals. The property also has an adequate road system and a set of cattle pens (corrals) with a windmill and stock tank (pond) as additional improvements. It appears there is also some bird hunting available on the land. The seller has agreed to carry a note with thirty percent down for ten years at an interest rate of ten percent. If you were able to extract this information from the abbreviated advertisement, you are to be congratulated! You are indeed an extraordinary sleuth, or else you possess the ability to translate hieroglyphics. This example could be considered overkill, but the point is important to remember. Practitioners need to make certain that they are not the only ones who can interpret the advertisement; otherwise, they may be the only ones who read the advertisement.

Also remember the tried and true acronym AIDA (Attention Interest Desire Action) as it relates to advertising. Attention comes first, then interest is generated, followed by desire, and hopefully action occurs. Well-written advertisements must have a specific heading in order to receive attention. Most advertisement writers spend too little time on this critical activity. It does not matter how good the information in the advertisement is if the heading does not catch the reader's eye. The body of the advertisement needs to address relevant details for the readers. The specifics would include the location (as general or particular as the practitioner deems appropriate), some (and not all) of the improvements to be found on the land, and possibly a reference to price. Whether or not to include price is a frequently debated topic in real estate circles. Some practitioners swear by putting a price in the advertisement while others only swear at the advertisements that provide all readers with the price. Opinions vary, but our preference is to place the list price per acre or total sales price being asked in advertisements. This practice helps when qualifying prospects who respond to the advertisement. Keep in mind that advertising is a costly expenditure in marketing real estate. All practitioners should monitor their methods carefully to ensure the effectiveness of their approach.

SHOWING PROPERTY

Years ago, there was a fuzzy white puppy that lived at a house in the area. The dog was not worth much in the way of usefulness, except that it thought its job was to protect the yard. Whenever the big garbage trucks came down the street, Kyp would become very agitated. The dog would then chase the trucks for a block or more, and then trot back home apparently satisfied that it had done its job. The family often wondered what would happen if Kyp ever caught one of the trucks. Much like Kyp, the agents may have been running around trying to snag something that they would not be able to handle if they were lucky enough to catch it. Now the chance has come to present a property. This part of the chapter reviews a few of the reasons **why** practitioners show properties. Later in this chapter, **how** to show properties will be discussed.

The Purpose of Showing Property

There are many reasons why practitioners show or demonstrate their listings. The most important reason is to **sell the property.** After reading this statement, you may ask why we would insult your intelligence by stating such an obvious point. No slur against anyone's intelligence is meant by this statement. But many times in showing property with other practitioners, we have found the other agent to be so very unprepared it is embarrassing for everyone. The best thing that can occur when showing a property is that the prospects like it and want more information about the property. Better yet, the prospects will want to make an offer for the property.

One of the worst things that can happen is that the agent is unable to respond to reasonable questions about the property. Or worse yet, the practitioner cannot properly prepare a correct and valid offer to purchase document. The best advice is not to go out to show a property unless you know enough about the property to be of assistance and service to the people being shown the farm or ranch. More importantly, you should not go even then if you are incapable of putting an offer from the prospect in the correct format should the prospect decide to make an offer. Be ready to show and do it properly!

Another reason practitioners show the property is to **further qualify** a prospective buyer. Viewing a property together can often provide both the practitioner and the potential buyer with a great deal of information. It can provide the buyer a perspective on whether or not the subject property will meet investment objectives. Everyone who reads this will probably recall an instance, or maybe many instances, in which they thought a particular product or service would meet their needs. Upon investigation, it was determined that what was perceived as a solution and what was in reality a solution were two different things. When showing a property, this need not be a problem unless the agent chooses to view it as a problem. If a particular piece of land does not meet a buyer's criteria, that is fine as long as after the showing the practitioner knows more about what the buyer **does desire** in a property. Remember, one of the most frustrating buyer types to work with is the person who likes everything, but will not commit to anything.

Another reason to show property is for **involvement.** Real estate is definitely an emotional type of investment for the vast majority of people participating in the transaction. So the astute marketer of rural properties will endeavor to create opportunities for prospective buyers to actually "feel" like a part of the land. This can be achieved by utilizing a wide variety of methods. The following is a favorite example from the life experiences of one practitioner. When dealing with a buyer who is interested in the property for recreational purposes because of an interest in wildlife, the buyer is invited to observe the early morning or late evening game movements. There is something about sitting on a ridge and watching the sun rise or set. Observing nature in its purest and simplest form often evokes strong feelings toward the land. The same could be true of arranging for the buyer to watch the crops being harvested on an improved farm, or to take a horseback ride over the land. Please recognize the risk of liability that must be considered when utilizing such outings. The point here is to help the potential buyer become excited and learn or be reminded of how incredible an experience it is to "feel" close to the land.

There may be other motives in showing property, but practitioners are encouraged to use their imagination when developing a style for showing properties that will make the visit to the property a positive experience. Knowing **why** practitioners show property is important in order to develop a solid set of operating guidelines. The next part of this chapter will review the **how** of showing property.

Showing Guidelines

The following material is an important part of our effort to assist practitioners. Please note that the heading mentions "guidelines" and not carved-in-stone rules. While we strongly believe in each and every one of these suggestions, we recognize the need for adaptability and flexibility based upon the conditions present at a particular time and circumstance. Thus the term, guidelines, appears to be appropriate.

The first guideline is to **preview the property.** It is surprising how often so-called "experienced practitioners" want to show listings without the benefit of a preview. Certainly there are times when this is necessary; however, such occurrences should be the rare exception and not a standard practice. It seems to us next to impossible for experienced practitioners to be able to create a successful showing situation when little if anything is known about the land to be presented. Effective showings start before the agent opens the car door for the prospective buyers, and long before the prospects arrive at the property. In the long run, practitioners simply cannot effectively sell what they do not know or understand. There have been tales that seem to prove this guideline wrong. In fact, the authors have even participated in sales in which the land was never shown to the buyer or the buyer was never met. But it should be emphasized that this type of transaction is a rarity and definitely not the norm. Previewing does not cost, it pays!

Another guideline is to **carefully plan the showing.** Showing land effectively is somewhat like producing a play. It takes practice and planning to do either one well. Give some thought to the route, timetables necessary for appointments, the things you want to ensure are pointed out, questions you want to ask, etc. All are important aspects of proper planning.

Planning also entails **determining the audience.** This means finding out who is going to be physically present for the showing in order to ensure that adequate vehicle space is available. It also requires having to do an initial qualifying. So often, property showings that should never have taken place have been experienced. The property fits so few, if any, of the buyer's criteria that the effort and time spent to create the showing was wasted. There is also the risk of alienating the buyer or the seller when such preliminary work is ignored or forgotten.

As a property showing actually begins, the practitioner is advised to **watch for reactions.** There may be some strong likes and dislikes that you will be able to "read" in buyers simply by being aware of their verbal and nonverbal signals. This is not to imply that you need to be hypersensitive, but you do need to pay respectful attention to the things being said, and in some cases, to the things that are not being said. Practitioners often think that a buyer will be excited about the same things that are important to them. Frequently this is not the case and you must pay attention to the buyers' needs and not your own preferences.

Another guideline that suggests the need to focus on the buyer is to **let the buyer set the pace** of the property showings. How long does it take to look at a piece of property? This question is impossible to answer quantitatively since it will be based upon many factors unique to each situation. However, it is possible to get a reasonable idea of proper pacing by allowing the buyer some leeway on how long to remain at any one spot, or on the property for that matter. Some people need a considerable amount of review time to reflect and consider. This type of buyer may require several showings with each of those showings being longer than the practitioner would personally prefer. At the other end of the spectrum is the person who seems to take an uncomfortably short period of time for such a large-scale decision as property selection. At least, it may seem that way to the practitioner. The point is that you cannot dictate another person's decision-making time frame. So it is smart not to even try!

Balancing the above suggestion is an important guideline under the label of **soliciting feedback.** If you are letting the buyers establish the pacing flow and are paying careful attention to their involvement and reactions to the property, it is much easier to make this request. When couched in terms like "Are you going to buy this place?" too much pressure is placed on the buyers. On the other hand, consider this request: "Please help me look at this property through your eyes and see the pluses and minuses from your point of view, Ms. Buyer." This conveys a sincere respect and need to know without adding undue pressure. By asking for this type of comparison and listening carefully, you can help make the selling task easier.

Of course, sometimes potential buyers truly do not know what they think just yet. Or, if there is more than one decision maker involved, the various decision makers may need some "talk time" to discuss and compare impressions. The astute marketer recognizes and respects this fact and **gives all of the decision makers time alone.** No good can come from attempting to force a decision or even a comment if the time is not right to do so. The successful salesperson will attempt to get feedback, but if no feedback is immediately received, the salesperson will not become frustrated or defensive. This brings to mind another most important guideline. When asking for comments from prospective buyers, if they happen to be somewhat critical, **do not defend the property.** Criticism is frequently a type of value judgment. Nice, big, adequate, sufficient, lousy, poor, or any other term that comes to mind whether good or bad, is how one person or group views a particular property at one point in time. With further education and contemplation, such a viewpoint may change, or it may remain the same forever. Either way, it is not smart for the practitioner to become personally affronted by criticism of the property. The singular exception to this guideline is when a potential purchaser has the **facts** wrong. Then correction by the salesperson needs to take place, but only as it relates to the actual realities and not personal preferences or opinions.

Personal sensitivity is the subject of the next guideline. When sellers take pride in what they have created and hear harsh words about the property, bad feelings can be evoked. When an agent is asking for comments about a place and the seller is present, buyers may be less than candid with their comments. This lack of honesty may not be because of a desire to deceive, but rather out of respect for the seller's feelings. However, if the prospective buyers have a problem or complaint about the property, the salesperson needs to know this as soon as possible. Because of this potential, we recommend as a guideline that you **keep the seller away from showings.** This may sometimes be rather difficult, but here is a strategy to consider. By asking the seller to accompany you during your preview, you can become the beneficiary of the seller's insight into the property. Once the seller tutors you, it is advisable for you to acknowledge your respect for the seller's superior awareness. In addition, it is important to know that you can call upon the seller for additional assistance. But in the best interests of the sale, in many cases you should ask to show the property without the seller's immediate involvement. If the seller refuses, at least you have tried. It is also important not to press the issue too hard for fear of alienation. But we believe it is best to keep the buyer and seller apart during these first stages of the sales process if possible.

As you consider the multitude of guidelines provided in this chapter, hopefully it will become obvious that there is much to remember and to include in each and every showing. This leads to one more suggestion, that you **limit the number of showings.** Please remember that your goal in showing a property is to obtain comparisons between properties and not to create confusion. After many years of farm and land brokerage, it is obvious to us that *properly qualified* prospects frequently look at fewer, but more relevant, places when considering an investment. Through the years, we've encountered many frustrated buyers and practitioners who are thoroughly confused because of the excessive number of properties presented during one showing period, or over a longer time frame. We are not recommending you limit showings to only "X" number of choices in an uncompromising, absolute manner. Rather, we are suggesting that you make certain you are not doing the buyer, seller, and yourself a true disservice by attempting to present as many places as can be "crammed" into a single day, weekend, etc. The point here is to show fewer properties in a better method. This is far better than showing more and more in the desperate hope that something, or anything, will please a buyer. In this case, keep in mind that "more ain't better!" A memory jogger to reinforce this idea is: Qualify better to show less to sell more.

The final guideline is applicable for a particular individual showing and for all showings in general as well. Remember that effective marketing, when demonstrating any product or service, requires that you **keep the showings positive.** It is wise for a practitioner to be pleasant, motivated, informed, and, in a word, professional in all showings. Keep in mind that sometimes successful showings are based not only on what has been offered, but also on how well the practitioner goes about the actual presentation. It is also smart to remember that much work and preparation go into developing an effective showing. Few, if any, practitioners can ever have enough opportunities to show good, well-priced properties to motivated, well-qualified buyers. So make the most of each and every one of those showings. We hope that these guidelines will help you to show property in a professional and effective manner.

7 Property Condition

In other chapters the importance of knowing about soil and water conditions and qualities is highlighted. It is just as important, and maybe even more so, to be knowledgeable about improvements on the land. We strongly counsel that you make a **diligent** search to find out as much as possible about the property. This includes not only its strengths, but its weaknesses and shortcomings as well. Some states now require that a licensee obtain from the seller a "property condition statement," or some similar term, as a part of the listing agreement. This document will serve as a record of the seller's disclosure of any and all known problems with the land or improvements on the land, and an example is shown in Fig. 5-2.

Licensees should not depend solely on the seller for such information, but should do some "homework" on their own. Furthermore, it is advised that you make a written record of any additional concerns or findings you discover that raise questions about property condition. Should there be any doubt about the condition of an improvement, for the protection of all parties involved an inspection should be made by a trained, unbiased third party. Having these inspections completed **before** any active marketing is begun can be a safeguard for both the owner and the licensee. Sellers are often reluctant to invest in such an expense in advance. The best advice is to proceed with caution in a sale if there is a doubt regarding the quality or condition of any improvement on the land. At the very least this inspection should be part of the contract that will be drawn when the offer is accepted. See Fig. 7-1 for an example of the form used regarding some of the possible inspections to be negotiated in the offer. At the other end of the spectrum, it should be stated in the sales agreement that no warranties are expressed or implied. Please note that this area may require some legal counsel.

Another area where additional care and attention are warranted is the representation of the number of acres in specific fields or pastures. Astute practitioners will justify such acreage statements based upon something other than what an owner or they themselves have "guesstimated" the acreage to be.

An illustration such as the one presented in Chapter 12 is a good example. This plan breaks various plots into reasonably accurate estimates and allows the practitioner to have a somewhat qualified source to quote when referencing size. Other common sources are field notes and plats from surveys. Experience indicates that rarely does an owner order a new survey before actually receiving an offer on the property, so be prepared to work from an "old" survey. Make certain, if possible, that no major changes have occurred to render the survey inaccurate. If you cannot confirm that the survey is reasonably accurate and current, we advise that you refrain from using the survey. When an old survey is used, be certain to acknowledge the source of any statements made about acreage or changes to the property. Even if there is little doubt about size, it may be wise to consider using the term "approximately." Some earnest money contracts state that the offer is based upon an approximate number of acres and allow for some range of

Figure 7-1

03-17-92

PROPERTY CONDITION ADDENDUM

PROMULGATED BY THE TEXAS REAL ESTATE COMMISSION (TREC)

ADDENDUM TO EARNEST MONEY CONTRACT BETWEEN THE UNDERSIGNED PARTIES CONCERNING THE PROPERTY AT ______________________________
(Street Address and City)

CHECK APPLICABLE BOXES:

☐ A. **TERMITES AND OTHER WOOD-DESTROYING INSECTS:** Buyer, at Buyer's expense (except at Seller's expense in VA transactions), may have the Property, including outbuildings, inspected by a Structural Pest Control Business Licensee. If the inspection indicates treatment or repairs are required, Buyer will furnish a written report to Seller from such Licensee within _________ days from the effective date of this contract, but no treatment or repairs will be required for fences, trees or shrubs. Buyer's failure to furnish such report to Seller within the time specified shall constitute a waiver of Buyer's right to any treatment and repairs.

☐ B. **INSPECTIONS:** Buyer, at Buyer's expense, may have any of the items designated below inspected by inspectors of Buyer's choice. The purpose of the inspection will be to inform the Buyer in a written report if the items inspected are functioning at the time of the inspection, or in need of repair. Failure of Buyer to furnish written inspection reports and to designate the repairs to which Buyer is entitled by this contract within the times specified below shall be deemed a waiver of Buyer's repair rights.

Within __________ days from the effective date of this contract, Buyer will furnish Seller written inspection reports for the items indicated below with a designation of items in need of repair.

1. ☐ BASIC FOUNDATION SYSTEMS ONLY.

2. ☐ BASIC STRUCTURAL: Includes and is limited to: foundation systems, roof components, walls, ceilings, floors, basement, water penetration, fireplace and chimney, stairs and landings, attached balconies, decks and porches, exterior doors, windows, and screens. Excludes: Cosmetic features.

3. ☐ BASIC MECHANICAL: Includes and is limited to: plumbing system and water heater, electrical system, all permanently installed heating, cooling and ventilation equipment and systems, all built in cooking equipment, dishwasher, disposer, exhaust fans, and trash compactor.

4. ☐ OTHER SYSTEMS AND ATTACHMENTS: ☐ pressure test of gas lines, ☐ built-in security equipment, ☐ fire detection equipment, ☐ outbuildings, ☐ water wells, ☐ septic systems, ☐ permanently installed outdoor cooking equipment, ☐ lawn and garden sprinkler systems, ☐ swimming pool and equipment, ☐ spa or hot tub and equipment, ☐ sauna or hydrotherapy equipment and related equipment and the following items: ______________________________
____
______________________________.

☐ C. **OTHER REPAIRS:** Seller shall make the following repairs in addition to those required as a result of the specified inspections:
____
______________________________.

All inspections shall be by persons who regularly provide such service and who are either licensed as inspectors by the State of Texas or otherwise permitted by law to perform inspections. Repairs shall be by trained and qualified persons who are, whenever possible, manufacturer-approved service persons and who are licensed or bonded whenever such license or bond is required by law. Seller shall permit access to the Property at any reasonable time for inspection or repairs and for reinspection after repairs have been completed. Seller shall only be responsible for treatment and repairs specifically designated or required by Buyer in accordance with this addendum, subject to the provisions of Paragraph 7 of this contract.

NOTE: SELECTION OF INSPECTORS AND REPAIR PERSONS IS THE RESPONSIBILITY OF THE PARTIES AND NOT THE BROKER OR SALES ASSOCIATE.

______________________ Buyer ______________________ Seller

______________________ Buyer ______________________ Seller

The form of this addendum has been approved by the Texas Real Estate Commission for use only with similarly approved or promulgated forms of contracts. No representation is made as to the legal validity or adequacy of any provision in any specific transactions. (03-92) TREC No. 2-3. This form replaces TREC No. 2-2. **001A**

№ 007

deviation, say ten percent. If a new survey reflects a deviation beyond that allowable range, then frequently the contract will become null and void, or will have to be renegotiated. Another way of getting fairly accurate acreage is with the use of a planimeter. The local FSA or NCRS office will usually be able to help practitioners calculate acreage. A word of caution: Be certain to get written authorization from the landowner that gives you permission to obtain the requested information. The wise practitioner will heed these warnings and be most careful when presenting the "facts."

FINANCING FARMS AND RANCHES

It certainly would be nice, and much easier, if every transaction were done on the basis of the buyer having adequate cash reserves on hand to pay for the property immediately at the time of title transfer. And while some sales actually occur in this manner, many farm and ranch sales require some method of financing. It is virtually impossible for anyone to know all the sources of money available throughout the nation. Even if a practitioner does have a handle on locating financing sources, circumstances could alter overnight in a particular area due to economic or legislative changes. Therefore, reference is made here in a general nature to some forms and sources of which practitioners should be aware. This section of the book will only serve as a starting point for a practitioner's particular needs and circumstances.

Before going into the various sources of financing, here are some general guidelines as to how loans on farms and ranches may differ from those on other real estate. These differences are sometimes a shock to the potential borrower or practitioner if they are inexperienced in this aspect of real estate. Extensive experience indicates that lenders tend to focus heavily on three factors when evaluating agricultural land. These factors are:

1. Land quality.
2. Borrower ability and experience.
3. Borrower financial strength.

Each of these three factors is worthy of some mention and explanation. As explained in other parts of this book, the shared information on quality of soils and other factors to consider in evaluating a farm or ranch are important to lenders who do this type of financing. Remember this most important fact: Most lenders tend to evaluate agricultural property on its **productive value.** So the speculator looking for future developmental potential or the recreational buyer looking for a "fun getaway place" may find the lender is very conservative in an opinion of value. After all, the lender may have to "eat" the loan if things do not go well for the borrower. Such a difference of opinion is easy to illustrate. Market value of recreational land in a popular area may be many times more than the property's worth based upon productivity. This is less of a problem when times are good and money is flowing. But what happens when the area economy falters and the "hot spot" is no longer so magical? Factors like this give loan officers insomnia. However, such concerns do bring out valid points from a prudent investment perspective. We recommend that potential buyers or borrowers ask this most relevant question about any property being considered: "What will this property do for me to earn its own keep?" Keep in mind that dramatic sums of money have been made in land investing. Unfortunately, equally large sums have been lost on the same type of property.

Another factor the lender will scrutinize is the experience a borrower or buyer brings into the investment. While it would be great to be twenty-five years old with twenty-six years of experience in dealing with the exact same type of land being purchased, this is not often the case. A successful background in one area is not an absolute guarantee of equal skill and ability in another, but it can help. This issue of experience becomes very important for "newcomers" to farm and ranch ownership. It is also looked into most carefully when a person with much success and background on inherited "family land," but none on "new mortgaged land," wants to borrow money for expansion. The practitioner will strive to impress the lender by emphasizing any and all of a buyer's experiences and successes that may make the difference between a favorable decision and a rejected loan application.

One of the most frequently considered aspects of any lending opportunity is that of the mortgagor's financial ability to repay the debt. A prudent review of certain areas is a normal and logical requirement of financial institutions or individuals when evaluating a borrower. Be prepared, or have the buyer prepared, to present proof to the lender of past income by virtue of copies of tax returns. Several years, typically three to five years, of the latest tax returns may be requested by the lender. Likewise, a financial statement reflecting all debts and other assets, both personal and business ventures, if any, should be prepared for the lender. Such documentation is often requested by lenders and it may also be necessary for these documents to be audited, confirmed, or certified by a professional such as a certified public accountant. In any case, such information must be current and accurate to be accepted by any lender. While these requirements are not unusual or unreasonable, such documentation may take a considerable amount of time to prepare for many buyers or borrowers. The practitioner should advise buyers and borrowers to be prepared to provide copies of their tax returns and current financial statements to the lender. A considerable amount of time can be saved by having these records in order prior to making an appointment with a loan officer. In general, the following has frequently been found to be a good idea: Have a buyer clarify with a lender (or lenders in some cases) what loan amount the buyer can qualify for from each institution before extensive showing, negotiating, or contract preparation begins. Everyone can then go forward with greater confidence and enthusiasm. Keep in mind that a purchaser needs to be financially able and not just ready and willing. No one "wins" when a sale falls apart due to a buyer's inability to obtain financing. And, if a licensee represents the seller in such a situation, hard feelings and a loss of confidence in the licensee could occur.

SOURCES OF LOANS

Within the immediate area of the subject property there may be **local lenders** who represent sources of financing. For land located within an economy that is predominantly agricultural, the loan officers may be very comfortable and experienced with farm and ranch land loans. In fact, these people may represent a valuable resource in terms of assisting a potential buyer to assess land values in the area as well as productivity considerations. In many cases the subject property may be one that the lender has already encountered. Short of firsthand knowledge of that particular tract, the lender may still be comfortable with the subject property based upon experience with similar properties.

A word of caution when dealing with locals is warranted if the practitioner is a "newcomer." It is important that the practitioner avoid the pitfall of debating or attempting to tell the lender, or other professionals, how to run their business. It is very easy for agents to become frustrated with business practices they may not be accustomed to or fully understand. To become critical of such practices will cause the agent much more harm than good. Become aware of the lender's rules and practices prior to applying for a loan, and then both the practitioner and the borrower may find dealings with the lender are more easily handled. From a purely practical point of view, if the agent is an outsider, it is best to simply try to learn the requirements and expectations of the lender. The practitioner should then respect the requirements if the loan source is of long-term importance. If borrowers can do better with another lender, they should simply go where they will receive the most beneficial loan terms. But it is important for practitioners to remember not to burn any bridges that may be needed farther down the road.

In some circumstances, another good source of loans for land is the **buyer's own bank.** Through the years many good properties have been sold to very successful businesspeople with well-established lines of credit already in place. If such circumstances exist, it would be wise to avoid "reinventing the wheel," especially if satisfactory terms can be arranged between a lender and borrower who already know one another. In cases like this, the licensee is frequently called upon to be the source of a good deal more information than in the above setting. The lender may need some guidance on values in the area, but be careful about trying to wear more than one hat in a single sale. While you may be able to provide some market data, we strongly caution against trying to be an unbiased appraiser while at the same time being involved in the transaction. It is much better to give a lender the names of some competent appraisers with specific experience in area farms and ranches. Then you can be a resource in every other possible way to help the appraiser if required. This allows you to maintain your objectivity and not create doubts about your professionalism.

Remember, it is a good idea to have some awareness as a borrower or practitioner trying to help a buyer find financing, as well as understanding of bank policies and practices in order to determine if a loan is even possible. The charters of certain lenders prevent them from going outside of a certain geographic region, for example, the county where the lender is located. Or there may be certain loans that are not allowed, such as for raw land, which is a most speculative asset. A borrower with other assets available for collateral may use acceptable alternative security and still obtain the money required.

It is well beyond the scope and intent of this book to offer a history of rural finance; however, there are some fundamentals that are worth discussing briefly in this chapter. In some parts of the nation a common source of money for agricultural properties is the **Federal Land Bank.** The Federal Land Bank is a part of the Farm Credit System that has been in existence for many years. The Farm Credit System is a nationwide financial cooperative that makes loans secured by rural land. Congress created the system in 1916 to provide agriculture with a dependable source of credit. More than one quarter of the credit needs of domestic agriculture are met by system institutions.

System banks and associations do not take deposits. Funds are raised through the sale of bonds in capital markets. The funds are loaned through a nationwide network of approximately 232 Farm Credit lending institutions. Typical services offered by the regional Farm Credit banks and associations to eligible borrowers include real estate loans, operating loans, and rural home mortgage loans.

The structure of the Farm Credit associations varies around the country. As of 1995, there were 72 Federal Land Bank Associations (FLBA), 69 Production Credit Associations (PCA), 60 Agricultural Credit Associations (ACA), and 31 Federal Land Credit Associations (FLCA). Long-term mortgage loans are typically made through FLBAs or FLCAs. Short-term and intermediate-term operating loans are generally made by PCAs. In some areas, these lending entities have combined into ACAs, which provide both long-term and short-term loans.

As a lending cooperative, Farm Credit today is wholly owned by its customers through their purchase of stock. Stockholders elect directors who manage the business. Money is loaned for the purchase of agricultural property, rural homes, or recreational properties. Money is also loaned to refinance and improve property. Mortgage loan terms range from five years to forty years. Options include fixed interest rates, variable interest rates, adjustable interest rates, and interim loans for construction. There are monthly, quarterly, semiannual, or annual payment options.

A real advantage to working with the Federal Land Bank is that loans can frequently be made for long periods of time, thirty to forty years in some cases. Other lenders cannot or will not usually agree to such a long-term loan for farm and ranch land.

One negative in dealing with the Federal Land Bank is that loans are made upon the Bank's viewpoint of value which must, by law, be based upon productive value. This frequently means a fairly low loan-to-value ratio when the market value exceeds production value.

An example of this dilemma may be helpful. A buyer agrees to pay $800 per acre for a tract of ranch land. The Federal Land Bank appraises the property's productive value at $600 per acre and is willing to loan 75 percent of that amount, or $450 per acre. Under this scenario, the buyer has to make a $350 per acre down payment, or approximately 43 percent! This can partially be overcome if the seller is able to offer some secondary financing, but this is not always a feasible alternative.

Another challenge practitioners often face when dealing with the offices within the Federal Land Bank system is the amount of processing time required. Because there are only a few hundred offices of the Federal Land Bank for the entire country, some of these offices cover large areas. The employees of the Federal Land Bank have large numbers of counties and loans to underwrite, process, and close. The time frame to obtain a loan may be too lengthy for some transactions. The borrower and practitioner would do well to remember that many Federal Land Bank loans are variable rate loans, which means the interest rate charged will change at specified intervals as described in the note. A hard-pressed farmer or rancher could receive a tremendous shock when the interest rate increases and their overall cost of money skyrockets.

Another association that can be a source of funds under certain conditions is the Production Credit Association (PCA). While the major focus of this cooperative is on short-term operational and equipment loans, the PCA does occasionally make loans using the real estate as collateral.

Any practitioner or potential buyer who plans to work in a particular marketplace would be well advised to make contact with the area Farm Credit System representatives. Learn all there is to know about what these entities can and cannot do as lenders. By the way, these organizations are sometimes required to be the seller when a default occurs and is followed by a foreclosure. This is another good reason for a practitioner to stop by and become acquainted.

Detailed information regarding the tax implications of selling property or owner financing of property are not within the purpose of this book. While no claims of providing tax expertise are made within this text, it must be emphasized that for many sellers, the benefits of financing the sale can far outweigh the negatives. For example, Herman and Jo Ann Seymour are selling their farm. After many years of hard work and sacrifice, Mr. and Mrs. Seymour now own the farm free and clear. For the couple to accept all cash from the sale would result in an enormous tax bite, and therefore, they would have less disposable income. Since the sale of their farm will be the main source of funds for the remainder of their lives, this is significant. Furthermore, it is unlikely that they would be able to receive as high an interest rate in a savings plan as they would on a first lien loan on the land.

To have more money (principal) available and a higher rate of return (interest) is the best of both worlds. Of course, there is a downside that must be considered. When a seller finances the sale of property, there is always the risk of default which will require foreclosure in order to protect the seller's interest. And, there are some management duties associated with the monitoring and collection of income.

These challenges can be somewhat moderated by carefully qualifying potential borrowers or by using the services of a trust officer or business manager to monitor the business interests. Years of experience have proven that many sellers only "give up" their land when health, energy, or aging issues force them to do so. At that point in time, there is often little interest on the part of sellers in continuing to keep books and monitor business practices. Their children are often far removed physically, lack the experience necessary to keep the farm going, or are simply too busy with their own business or profession to take on the extra burden of managing a farm.

Here is one more note worthy of serious consideration regarding the seller's decision about financing the sale for the buyer. Frequently, a seller has used a particular accountant, attorney, or bookkeeper for decades. These people may not be the best sources of advice for the seller regarding the implications of a sale or financing program. It is important to remain calm and not get excited about this statement. After all, this statement is not a put-down for older farmers or ranchers, small towns, or country folks. The intent here is to explain that sometimes it is appropriate to obtain a second opinion, especially from an outsider who does not have long-term relationships with the people involved in the transaction. A practitioner who is involved in a sale owes it to the client to make this suggestion. Whether or not the seller follows up on the suggestion is something that is outside of the control of the practitioner. There is only so much that can be done for someone else. Remember that no person can be all things to another person in business or otherwise. Know your limits and act accordingly.

CONTRACT PREPARATION

There is never a better time to remember your limits than when the time arrives to actually draw up a sales contract. There are many variations in the terms used around the country so it is important to remember to call this very important document by the name used in your area. Earnest money contract, sales agreement, and purchase contract are only a few of the numerous names used to identify the contract. No matter what the contract is called, the document is an agreement between a buyer and a seller. The contract also outlines the terms and conditions of a potential sale. The word *potential* is emphasized because until both parties have come to an agreement on all factors involved with the transaction, the contract is not a legally binding document. Furthermore, most contracts require any number of conditions to be met before the actual transfer of title can occur. Nonetheless, the sales contract is by far the most important document required in a sales transaction. This is because the contract "sets the rules" for the actual sale transaction.

Given its importance, the astute buyer, seller, or salesperson will want this document to be very carefully prepared. At this critical point in the transaction, the practitioner must remember not to "fake it until you make it." Seeking the advice of or review and approval from the principal's attorney, an experienced broker, or anyone else experienced with contracts is not a waste of time. It may, in fact, be what cinches the deal. However, we do not advocate allowing too many people to become involved during the negotiation process. Time and time again, excited buyers, sellers, and practitioners in a hurry create an inappropriate contract that does not reflect the intent of the parties, or draw the contract improperly, which makes it useless. Contract creation is not a process to be rushed through. If you prepare the contract properly, the final stages of the sale as presented in the next section will be easier for everyone involved.

8 Finalizing the Sale

In the previous material, we emphasized the importance of creating the proper environment of information and accuracy to obtain an offer on the property that would satisfy both the seller and the buyer. In this chapter, the steps needed to ensure a smooth closing or consummation of the transaction will be reviewed. A perspective on the critical stage of title transfer is provided, and several specific suggestions on how to "keep the sale together" are discussed.

AGREEMENT IS NOT CONSUMMATION

Often, an agent wants to relax once the difficult phases of offers and counters are completed. This tendency is understandable even though it is unwise. It is important to remember that all the practitioner has at the moment is an *agreement* to buy and sell. There are still a great many "hoops" for the agent to jump through before this sale is closed. To become overconfident or allow any other involved party to do so at this time is a grave error. There is much that needs to be done before you can "let the party begin."

Some of the tasks that still need to be accomplished are discussed here. These recommendations are in no way a complete list and may vary in different parts of the nation. This checklist simply represents a starting point for the practitioner. While each transaction is unique and will require some special action, there are some tasks common to all sales.

COUNTDOWN TO CLOSING

Review the contracts again. It may seem redundant to once again make certain that you are aware of exactly what the documents between the buyer and seller require each party to accomplish.

Think long and hard about what one veteran farm and ranch broker had to say about this most important time of a sale. Her wise comment was that "Relationships between all parties are never better than just after agreement has been reached." It is true that usually everyone is excited and anxious to get things done initially. Once the reality of the tasks ahead becomes more evident, buyers and sellers often lose some of their motivation. The astute practitioner will make certain that everyone knows their respective tasks and will assist all parties in moving forward toward task completion.

Process the "paperwork." Once the contract, binder, or sales agreement (or whatever name the documents are called in your area) has been completed, some paperwork processing must be done. Frequently, the agent involved will need to deposit any earnest money with the agreed upon escrow officer. Again, remember that practices vary from state to state and region to region. Certain states require the licensee to hold the money in a special account. No matter who holds the earnest money, the funds must be deposited. There is no legally binding contract in some states until the earnest money check has cleared the bank.

All interested parties will want to have a copy of the contract. Since many sales have several offers and counters as negotiations go back and forth, practitioners should make certain that they have the final offer that was accepted to present to the principals involved, their respective attorneys, any title or abstract companies, lending institutions, appraisers, and whoever else may need a copy. Practitioners should not fool themselves by thinking this is a low priority. *Properly* and *promptly* handling this significant document is a most worthwhile activity.

KEEP EVERYONE INFORMED

Another very important task for the agent is to keep all the involved and interested parties current on the progress of the sale. A practitioner can truly "keep the customer sold" by being excited and motivated as well as providing reports on the actual movement toward title transfer. One technique that has been used with success is to set a certain time aside each and every week to update all parties on the status of each sale. Even if there is little or nothing to report, this contact will keep the sale fresh and important in everyone's mind. A word of caution though: It is important that the practitioner not become a "pest" to lenders, escrow officers, appraisers, or other professionals involved in the sale. You do not want to wear out your welcome. After all, your sale is not likely to be the only sale on which these important resource people are working. If you become demanding, remember that any one of the above-referenced people can have a big impact on the ease or difficulty of the transaction. Give all parties credit for what they do to help finalize the sale. And be careful not to overstep the boundaries in terms of what the practitioner's expectations are.

KEEP THE SALE MOVING

The suggestion to keep the sale moving implies that no agent should become passive or disinterested in the sale progress. There is a very fine line between being persistent and being pushy. Certainly you will want to keep things moving ahead and sometimes this means that you will have to "encourage" different involved parties to do their part. For example, frequently sellers procrastinate about having certain repairs made that are a pivotal part of the sales agreement. Or, the buyer does not get all the financial records put together for the bank. Every once in a while, an attorney will delay in creating or reviewing necessary documents. Such minor obstacles are irritating and frustrating, but these obstacles must be viewed as a normal part of the latter stages of a sale. At least one party must wear the bad guy's black hat and keep the details of the sale moving forward. One way to eliminate some of the problems that go along with this aspect of the deal is to create for the buyer and seller a list of all the items they need to accomplish. As each task is completed, both the practitioner and the others involved can check the item off the list. This visual aid reminds the principals of their duties and also serves as a progress report for the busy agent with multiple transactions in different stages of development.

There are some particular areas that have been encountered or observed on a routine basis that deserve special mention. One common trouble spot centers on the various government programs with which a property may be involved. It is necessary to clarify and follow the appropriate process for transferring these programs to the name of the buyer in a prompt manner. Some programs allow for assumption by a new landowner, while others call for new applications, and still others have rather short time frames for change. Such details may seem tedious, but when there are subsidies, guarantees, or allotments involved, these all translate into money for someone. To that someone, attention to detail is important and necessary.

Another consistent area of trouble centers around leases to third parties. Some states require a new landowner to honor existing leases for the duration of the lease term unless there is an agreement that states otherwise in the original lease. Other states require that the lease be automatically terminated when a sale occurs. Practitioners must know what their state laws have to say about this issue. There must be clarity among the parties as to what the lease agreement (hopefully one in writing and signed by the landlord and tenant) states regarding a sale situation. Also, make certain all impacted parties, such as a tenant or subtenant, are fully informed of the progress of the title transfer.

One more troublesome issue is the debts and income associated with unharvested crops or produce at the time of title transfer. There are many ways to handle this situation, and we do not offer any one method as the best way to operate. However, the necessity of obtaining clarity among the parties on this important point must be emphasized.

PREPARE PEOPLE TO CLOSE

This idea may come as a surprise suggestion to some readers, but it is a most important step for the practitioner to take while planning to close a transaction. Helping people to prepare to close means several different things. One is confirming that all necessary paperwork such as notes, deeds, and title related documents are correctly prepared and available at the proper time and place. This step may require "delivery services" on the part of the practitioner involved in a transaction, i.e., carrying things around to the various parties to review or approve. This is not a demeaning position, just a very essential one. Another part of preparing to close is making certain that any questions the principals may still have are addressed. Often new questions will arise from a buyer or seller as actual consummation of the transaction nears. Practitioners are counseled to ask all principals if they have any questions, or if they need clarification on any point. To ensure this is emphasized, the licensee might utilize the strategy described below.

The day before closing is scheduled to occur, make a call or arrange to see in person the principals involved with the sale. Point out that closings can be rather simple and easy finalizations **if** there are no negative surprises or uncertainties for the parties involved. Inquire if you can be of service to them with any last-minute tasks or answer any questions. By creating an environment in which anything on the principal's mind can be addressed, you have done your part to ensure an opportunity for clarification exists. There is no absolute guarantee that all of the questions asked will be easily answered, but you will have done your best. Be careful about overstepping your role. If legal or financial questions are brought forth, it is best to refer such questions to an attorney or banker for clarification. If other licensees are involved, do not "go around" them by directly contacting their principal without informing them about the contact. For good future relations, it is best, in most cases, to communicate through the other licensee involved.

Still another part of preparing to close is to confirm closing costs for the buyer and seller. Determining how much money the buyer must arrange to bring and what the seller can expect to receive are simple but very important details. Earlier in the sale, at the time of negotiations for example, the principals should have received an estimate of closing costs. But now is the time to be as exact as possible and to be able to explain any difference between actual and estimated amounts.

There is one more point to emphasize about preparing to close. It is very strongly advised that **anything** the broker/agent becomes aware of that might create a problem or prevent a smooth closing should be addressed long before the scheduled appointment arrives. There are many horror stories about principals arriving to consummate a transaction and being caught by surprise by something that confuses or frustrates them. When such an experience happens, sales that **should** have and **could** have closed fall apart. If there is a problem, as soon as you become aware of it, you should make certain you understand as much as possible about the problem. Next, you should review solutions that you objectively consider reasonable and give the solutions a priority ranking in your mind. Then you need to contact your principal as soon as reasonably possible. Too often, the temptation is to do nothing and hope some miracle will resolve the situation at the closing table. That is poor reasoning on the part of a trained practitioner and it can also place a great burden on the escrow officer handling the closing. If you use the "bury your head in the sand and hope it all works out" approach, a few deals may be saved in spite of your best efforts. However, you run a very big risk of alienating the principals and other support people involved with the transaction. Practitioners should handle problems and not avoid having to deal with them. It is your professional responsibility and you owe such effort in your agency relationships.

THE ACTUAL CLOSING EXPERIENCE

The time to close the transaction has finally arrived! Buyers and sellers are scheduled to arrive at their appointed times. Taking into consideration all of the guidelines previously discussed, there are still a few ideas unique to this particular point of the sale. It is advisable that you take your buyer or seller to the closing. Unless you are working with extremely punctual people who know exactly where to go for the "signing of the papers," you should volunteer to pick up your party and escort them personally to the closing. This travel time will allow you once again to answer any questions that may surface. It will also give you some time to help calm any fears that may have arisen. It is very common for a person to have a period of self-doubt just before such an important experience. This has been labeled cognitive dissonance by psychologists and is often referred to as "cold feet" by the general public. The astute practitioner will point out how normal such feelings are during important events. If you do not transport your clients to the closing, you should at least meet them at the closing site.

Upon arrival at the appropriate office, experienced agents recognize their role at this point in time and maintain a low and quietly supportive profile. The escrow officer or closing agent usually has a definite routine and method of handling the actual transfer of title. It is not proper nor will it be well received for the practitioner to show off and take the spotlight at this juncture. Practitioners are strongly advised to allow the person responsible for the closing to be fully in charge. You still have some very important tasks that will soon surface!

THE POST-CLOSING RITUAL

Congratulations! The sale is now completed. But this does not mean that the practitioner's work is finished. The moment immediately after all papers are signed is a perfect time for you to carry out three very valuable tasks. Consider these circumstances when the sale has just been finalized. Most of the time both buyers and sellers are feeling a sense of relief and gratification that the transaction is completed. You should capitalize on this magical moment and do the following:

1. **Congratulate** the buyer and seller on the accomplishment. After all, you could not have completed the sale without them!

2. **Thank** the buyer and seller for their confidence, help, and support.
3. **Prospect** for future business. Keep in mind that there is no better promotion than a satisfied customer.

Ask to do business with these people again and request that they endorse you to their contacts. The sale should not be considered complete until you do these three things. You will then be able to begin again on a positive, upbeat scale, more confidently seeking to begin another sale opportunity.

There may be many other things involved in actually bringing a particular sale to the closing table and making it a "done deal." However, if you incorporate the ideas presented here into the last steps of the sales process, we believe that you will not only **make** more sales, but that you will be able to **keep** more sales together until final consummation of the transaction.

Being a sales leader is based upon completed business. This is not horseshoes, and simply being close does not count. The true professional in the farm and ranch brokerage business is the man or woman who can understand the need for technical skills combined with people skills and then incorporate a healthy dose of self-discipline. When this combination is capped by a desire to serve those buyers and sellers who come forth in an honest manner with caring and integrity, long-term success is an exciting forecast.

III

UNDERSTANDING THE EVALUATION PROCESS

9 Appraising Rural Property—The Process

This chapter describes a general approach to the appraisal of rural property. This material is intended to help the real estate practitioner in the marketing of rural property. Before suggesting a "list price" (asking price) for a property, a real estate broker or agent needs to estimate the market value or market value range of the subject property. Typically, the estimated value will be increased five percent to fifteen percent to establish a list price. (This is done in anticipation of the bargaining process that will normally precede the sale of the property.) In any case, **market value must be estimated** in the beginning stages of marketing a property.

A market value estimate made by a broker is often called a "broker's opinion" and by definition is a product of an appraisal. An appraisal may be defined as "the act or process of estimating value; an estimate of value."[1] Typically, we think of an appraisal as being a lengthy, expensive, typed document that is the product of a professional appraiser. This is certainly not always the case. Appraisals may be oral or written. "The length, type, and content of appraisal reports are dictated by the client, regulatory requirements, the courts, the type of property being appraised and the nature of the problem."[2] State Certified Appraisers have strict appraisal guidelines that must be followed. These guidelines do not apply to brokers and agents estimating market value for the ultimate purpose of marketing a property. In any case, a credible appraisal must be based upon current market data and sound reasoning and analysis.

It is true that appraising is not a science. There are few absolutes regarding the appraisal of rural properties. Good judgement, common sense, and experience are key factors in determining accurate value estimates. Certain appraisal techniques and methodology will be offered in the following pages. However, it should be remembered that different appraisers may utilize different appraisal techniques and methods. Often there are numerous ways to accomplish the same goal.

In the appraisal of real estate, there are three approaches to value, the cost, sales comparison, and income approaches. It is the opinion of many appraisers that the most reliable method of estimating the value of rural property is the sales comparison approach. This is because the sales comparison approach is the method that most closely reflects the thinking of buyers and sellers. In other words, this is the method of estimating value most used by the market participants, i.e., buyers and sellers.

The income approach may be applicable in the appraisal of specialized property that generates significant amounts of income such as dairies, feed lots, irrigated farmland, etc. But for many farm and ranch properties, income generation is not a factor considered by buyers and sellers in the estimation of value.

For example, recently a five hundred acre tract of land in south Texas sold for $550 per acre. The property was entirely rangeland with a cover of native brush. The land was leased for $12 per acre for hunting and livestock grazing. Ownership expenses, including property taxes and repairs and maintenance, were $1 per acre. The net operating income before depreciation and interest was $11 per acre. The capitalization rate is calculated to be two percent ($11 divided by $550 = 0.02). Obviously, the income generated by this property is not significant. When interviewed, the buyer of the property indicated that he did not utilize the income approach in his estimate of value of the property.

The cost approach is useful in estimating the market value of a property with structural improvements, i.e., barns, houses, etc. The problem with the cost approach is that it is very difficult to identify the various forms of accrued depreciation which include physical, functional, and external obsolescence. This is especially difficult when the improvements are older and have minimal contributory value. The sales comparison approach provides a more direct way of measuring the contributory value of the improvements and does not require identification of each individual form of depreciation. This method will be covered later in the material.

Since this chapter is not intended to be a complete appraisal guide for professional appraisers, the sales comparison approach to value will be emphasized. The sales comparison approach is "a set of procedures in which a value indication is derived by comparing the property being appraised to similar properties that have been sold recently,[3] ... The sales comparison approach is based on the premise that the market value of a property is directly related to the prices of comparable competitive properties."[4]

Measuring the minds of buyers and sellers is the goal of an astute appraiser. Appraisers must answer questions such as: What is the effect on value of paved road frontage versus no road frontage? What is the effect on value of having all of the mineral rights versus none of the mineral rights? These questions are answered by examining sales of rural properties that are processed in the sales comparison approach. Sale prices reflect the thinking of buyers and sellers and their **reaction** to road frontage, mineral rights, etc.

Appraisal reporting is not discussed in this material. It is important for the real estate practitioner to estimate value in order to determine an appropriate list price or asking price. It is the goal of the agent or broker to sell the property, not prepare an appraisal report. In some states it is required that a written disclaimer be provided to the client stating that the broker's opinion of value is not an appraisal and does not conform to the Uniform Standards of Professional Appraisal Practice. Care should be taken to conform to any regulations.

When appraising a tract of agricultural property, the real estate practitioner should:

1. Identify the purpose and function of the appraisal.
2. Identify and obtain relevant information regarding the subject property from the property owner.
3. Inspect the subject property.
4. Determine the highest and best use of the property.
5. Obtain comparable sales.
6. Process the sales comparison approach.

The purpose and function of the appraisal (as it pertains to this discussion) is to estimate the **market value** of the subject property as of a current date in order to set a listing price. Market value may be defined as follows: "The most probable price, as of a specified date, in cash, or in terms equivalent to cash, or in other precisely revealed terms for which the specified property rights should sell after reasonable exposure in a competitive market under all conditions requisite to a fair sale, with the buyer and seller each acting prudently, knowledgeably, and for self-interest, and assuming that neither is under duress."[5]

The definition of market value helps clarify the purpose of the appraisal. It says the value estimate to be identified represents the price that would be paid by a typical buyer (not a relative or a neighboring property owner), in cash (or its equivalent), after the property has been marketed in a competent fashion. Typically, market value does not involve owner financing, sales to relatives, sales made by parties in financial distress, or any condition not considered to be common and ordinary. These criteria are very helpful in selecting comparable sales. The sales price of comparable sales used in the appraisal should represent market value. Market value should not be confused with replacement value, investment value, or numerous other values.

After determining the purpose and function of the appraisal, specific detailed information about the subject property must be collected. This information is generally obtained from the property owner. A list of some of the important items that should be obtained is as follows:

1. Legal description and survey if available (including the number of acres in the subject property).
2. Mineral rights, production information, and royalty income, if any.
3. Outstanding mineral leases, grazing leases, hunting leases, etc.
4. Proof of legal access to the subject property.
5. Ingress/egress easements on the subject property.
6. Previous sales and listing information regarding the subject property.
7. Property taxes pertaining to the subject property.

A legal description is the address of a rural property. A legal description identifies a property as a unique parcel of real estate. It should indicate the surveys and abstracts or lots, blocks, and subdivision in which the property is located. The legal description should indicate the number of acres contained in a property. From the legal description, the property boundaries and therefore the shape of a property can be determined. This is important since property shape can affect utility and value. A long narrow property may be less valuable than one with a rectangular shape. The property owner may have a survey of the property, which greatly facilitates the appraisal.

In mineral active areas of the United States, mineral rights have a significant impact on value. In order to estimate the market value of a rural property, it is important to determine what mineral rights will be conveyed. If there is existing mineral production on a property, it is important to identify whether or not any production will be conveyed in the sale. If so, historical data regarding production and income should be obtained. Often, properties with oil and gas production sell with mineral rights, but exclusive of existing production.

Existing leases may affect the value of a property. For example, there may be an outstanding long-term hunting lease that encumbers a property. Such a lease would adversely affect the marketability and value of the property since the purchaser could not enjoy full use of the land.

If a property does not adjoin a public road, access must be addressed. If access consists of a right-of-way easement, a legal description of the easement should be obtained. If a railroad track must be crossed to access a property, a permit or easement should be obtained from the railroad company. In any event, the legality of access should be determined.

Neighboring landowners may have ingress/egress easements on (through) a subject property in order to access their property. Such easements may adversely affect the value of the subject property by limiting privacy. An ingress/egress easement located through the center of a property and through prime wildlife habitat or a farming operation would impair value.

Previous sales and listings for sale of the subject property may be helpful in estimating market value. For example, assume that a property has been listed for sale during the past several years for \$1,500 per acre. Also assume that the property has received adequate exposure in a competitive market. Common sense would indicate that the market value of the property is less than \$1,500 per acre.

The real estate practitioner should find the amount of property taxes pertaining to a property and also determine the paid status of the property taxes. Property taxes may be significantly different in adjoining counties. For example, in one county during the 1980's, property taxes were more than five times greater than in a neighboring county. Needless to say, property taxes were an obstacle to the sale of property in the "high tax" county.

The NCRS may have produced a conservation plan on the subject property. Be sure to ask the property owner because such plans contain aerial photographs with the location of stock tanks, water wells, pivot systems, and other items regarding the property. Such plans may inventory the existing vegetation (including specific grasses) as well as range sites and soil types. All of this information will help the agent or broker become more knowledgeable about the property.

After all pertinent information regarding the subject property has been collected, an inspection of the property should be made.

Preferably, the owner will provide a tour. It is desirable to drive the property boundaries and then make an inspection of each of the major pastures. A hand-drawn map of the property may be made during the inspection to record public road frontage, physical features, location of improvements, etc.

An inventory of each pasture should be made indicating the following information:

1. Topography—level, gently rolling, hilly, etc.
2. Soils—sand, silt, clay, sandy loam, etc.
3. Vegetation—native brush, cleared land with scattered trees and native grasses, improved pasture with permanent grasses.
4. Water sources—wells, stock tanks, creeks, river frontage, etc.
5. Fencing—type of construction, quality, and condition.
6. Livestock pens—description of construction, quality, and condition.
7. Oil and gas production, tank batteries, compressor stations, refinery, etc.
8. Adequacy of interior roads.
9. Underground petroleum pipelines, surface pipe, electric transmission lines, etc.
10. Structural improvements—ranch house, barns, sheds, etc.
11. Dump sites, gravel pits, dipping vats.

Dipping vats should be examined for the presence of toxic chemicals. Conditions such as dead spots or absence of vegetation in the vicinity should be noted.

Topographical maps and aerial maps are especially helpful in appraising larger tracts. For large tracts, it would be difficult to find important stock tanks, water wells, and improvements without the aid of a map.

Soil maps of the subject property may be obtained from the county NCRS. The pertinent information provided by the NCRS can be applied in managing farms and ranches. Information can be obtained for judging the suitability of tracts of land for farming, ranching, and wildlife. Soil survey books have been produced for some states in whole or in part. In these books are entire sections on use of the soils for wildlife, use of the soils for recreation, use of the soils for range, etc. The information that can be obtained from the NCRS is a tremendous resource.

An inventory of the listed items is suggested because these items have been demonstrated to have an effect on value. It is important to note the current use of the property. If a property is used for beef cattle production, note the current stocking rate (animal units per acre). If a game management program has been in effect, obtain details regarding wildlife numbers.

Inspection of structural improvements may be necessary when the improvements have contributory value. The contributory value of improvements should be determined from an analysis of comparable sales. Contributory value seldom equals cost, even of new improvements. (A discussion regarding this topic will follow in a later portion of this chapter.) An inspection of a rural residence may require measuring the exterior wall dimensions in order to calculate the gross living area (usually in terms of square feet). Notes should be made regarding the type, quality, construction materials, physical condition, and year of construction of the residence. Notes should begin with the foundation type and proceed upward through the exterior wall materials, number of stories, roof covering, and air conditioning and heating if any. An inspection of the interior should indicate number of bedrooms and bathrooms, living areas, floor and wall coverings, kitchen appliances, and condition. This information will be necessary when comparing the subject residence with the comparable sales.

Barns and similar buildings should be noted when these are estimated to have contributory value. A barn or shed may have utility, but no contributory value. The correlation between replacement cost and value may be slight. Contributory value can only be determined by analysis of comparable sales. Notes regarding barns should indicate the construction materials, condition, and size (square feet).

After a thorough inspection of the subject property has been completed, an inspection of the neighborhood is necessary. The land uses of the adjoining properties should be noted as well as other land uses in the area. The agent should note any urban influence in the vicinity as indicated by residential subdivisions. Any nearby commercial or industrial activity may be significant and affect value.

Area land use is important in determining the "highest and best use" of the subject property. Simply put, the highest and best use of the subject is the most reasonable and probable use that supports the highest value of the subject land. (This appraisal concept is useful later in this chapter as it provides a guide for selecting similar sales for comparison with the subject property.) Determining the highest and best use of most rural property is not difficult and area land use is an important factor in making that determination. For example, assume the subject property is a tract of ranch land currently used for beef cattle production and for hunting. Further assume that the neighboring properties also consist of ranch land as does the majority of the property within a ten-mile radius. The easy and obvious conclusion is that the highest and best use of the subject is for ranching and hunting. A textbook highest and best use analysis will require a more detailed approach, which considers the property "as improved," and separately "as though vacant," and requires four tests: Are alternative property uses:

1. Legally permissible?
2. Physically possible?
3. Financially feasible?
4. Maximally productive?

A detailed discussion of highest and best use is beyond the scope of this chapter. *The Appraisal of Real Estate,* Tenth Edition may be referenced for an in-depth discussion of highest and best use. For example, from a practical standpoint, the highest and best use of many rural properties can be broadly categorized as:

1. Agricultural—farming and ranching.
2. Recreational—hunting, fishing, etc.
3. A combination of both agricultural and recreational.

After the highest and best use of the subject property is determined, the search for comparable sales can be initiated. Appraisal theory says the comparable sales must have the same highest and best use as the subject property. For example, if the subject property does not have subdivision potential then neither should the comparable sales.

The quality of the comparable sales used in a rural property appraisal determines the accuracy of the value conclusion. Quality is determined by the similarity to subject, date of sale, arm's-length transaction, etc. The results of the sales comparison approach are only as good as the comparable sales used. That being the case, maximum effort should be spent in researching comparable transactions. Comparable sales may be obtained from the following sources:

1. County deed records
2. Local appraisers
3. County appraisal districts
4. Real estate brokers and sales agents
5. Bankers
6. Surveyors

In most cases, individuals involved in the real estate market (appraisers, brokers, etc.) are very helpful. The general attitude is one of cooperation so sales data are usually available. Once found, sales data should be verified with the buyer and seller of the transaction. Other parties involved (brokers and sales agents) should also be contacted for verification. Finally, the sales must be verified in the courthouse deed records so that the appraising agent can be absolutely certain that the transaction occurred. After the sales transactions are identified and verified, legal descriptions can be obtained from the warranty deeds. From the legal descriptions, comparable sales can be located on an ownership map (also located at the courthouse) and then located on county road maps showing highways, county roads, and cities.

If three or four truly comparable sales can be identified, ninety percent of the appraisal work is complete. In order to be good comparables, the sales should have the same highest and best use as the subject property and also have the following attributes:

1. Sales should be reasonably current.
2. Cash to seller transactions (preferred to owner-financed transactions and assumptions of existing notes).
3. Arm's-length transaction (not a distress sale or block-up sale).
4. Similar ownership rights or property rights (fee simple versus leasehold; and mineral rights versus no mineral rights).

The criteria listed above represent the best possible scenario for comparable sales. Often appraisals must be done in areas of limited sales activity. In these cases, less than perfect comparable sales must be used. For these sales, "adjustments" must be made to offset "atypical" characteristics (owner financing, distress sale, etc.) affecting the sale. (Characteristics affecting the sales price are referred to as **elements of comparison** and are discussed later in this chapter.)

When referring to a current or timely sale, the time that has elapsed between the date of sale (closing date) and the appraisal date is what is being addressed. What constitutes a current sale? The answer to this question depends on the marketing activity in the neighborhood. If the subject is located in a remote area and the appraisal is during a depressed market period, a current sale may be over twelve months in age. Typically, the market demand for rural properties (especially larger tracts) is small relative to the market for single family residences in major metropolitan areas. The market is small for large ranches because there are fewer people with sufficient income or wealth to purchase such properties. In comparison, there are many people with sufficient income available to purchase a primary residence in a metropolitan area. Therefore, the conclusion is that it is acceptable to use sales over one year in age in the appraisal of many rural properties.

Owner-financed sales should be avoided (when possible) since they are typically less reliable indicators of market value than cash sales. This is also true of sales involving assumptions of existing notes. Often the terms of owner-financed notes are advantageous to the purchaser and therefore a premium is paid for the rural property. Cash sales include transactions financed by third parties, i.e., banks, Federal Land Bank Associations, insurance companies, and savings and loan institutions. The result of the third party financing is that the seller of the property receives all cash and therefore the sale transaction presents a clear indication of market value. This assumes the seller paid no discount points. It is unusual for seller points to be paid on the sale of rural property.

"Conditions of sale" refers to whether or not a transaction was made on an arm's-length basis. An arm's-length transaction is defined as "a transaction between unrelated parties under no duress."[6] Arm's-length transactions generally exclude foreclosure sales, some estate sales, sales to adjoining landowners (which are also known as block-up sales), and others. Sales that are not on an arm's-length basis often result in prices above or below market value.

The subject property and comparable sales should have the same ownership or property rights. If the subject property is unencumbered by any long-term leases, then obviously the comparable sales should also be unencumbered. Using a comparable sale that is encumbered by a five-year hunting lease would result in a distorted value indication. A long-term hunting lease would dramatically impair the marketability and value of a rural property in many areas. Occasionally, a property will sell subject to a life estate. A life estate is defined as "total rights of use, occupancy, and control, limited to the lifetime of a designated party, often known as the life tenant."[7] A life estate is similar to a long-term lease. This arrangement would diminish the value of a property since the land or a portion of the land would be under the control of the life tenant. In some cases, the property may be shared with the life tenant. In any event, the sale price of a rural property encumbered by a life estate would probably be below market value.

As previously stated, mineral rights are property rights and have an important influence on value. Mineral rights typically influence value for two reasons:

1. Potential mineral income.
2. Control of the surface of a tract of land.

The contributory value of mineral rights of a rural property may be different from the market value of the mineral rights severed from the land. Owning the mineral rights with a property gives the owner control of the surface of the land. If a property is purchased without the mineral rights, as is common, the owner is somewhat subject to the control of a mining company (oil company) which may subsequently lease the property for mineral exploration and production. The owner of the surface will not have control over where or when a well is drilled. The owner may not be able to prevent an oil company from damaging the surface. Certainly the oil company could be sued for damages; however, if they are undercapitalized, there would be no point to a lawsuit.

An example of one such instance occurred in Frio County in south Texas several years ago when the surface owner of a small tract of cleared land (referred to as the subject property) contracted with the Farm Service Agency to lease the land under a federal subsidy program. (The lease required the landowner to plant the property in an improved grass and to leave it idle for the term of the lease. The intent of the program is to reduce soil erosion of highly erodible cropland. The government pays the landowner a handsome rental for participation in the program.) The subject property had been under lease in this program for several years when an oil company came onto the property, cleared a well site in the center of a field and destroyed the grass on the path to the well site. The landowner was not able to prevent the oil company from drilling since he owned none of the mineral rights. The oil company had a legal lease for the purpose of drilling a well. The actions of the oil company forced the subject property into noncompliance status with regard to the federal lease program since a portion of the improved grass was removed. The landowner was required to repay the federal government a portion of the lease money for the current year as well as past years for violation of the lease. Adding insult to injury, the oil company destroyed a portion of the fencing and ranch roads on the property and did not repair the damage. The landowner retained an attorney to investigate the case. The attorney recommended no action on the part of the landowner since the oil company was a small, undercapitalized company that was already involved in numerous other similar lawsuits. In other words, the oil company had no money.

For these reasons, some buyers refuse to consider the purchase of land without mineral rights. However, properties are bought and sold without mineral rights. When appraising land with no mineral rights, it is best to compare the subject property with similar comparable sales having no mineral rights. If the subject property must be compared to a property that does have mineral rights, an adjustment should be made. This adjustment must be extracted from comparable sales. This procedure will be discussed in Chapter 10.

There are other factors to consider when choosing comparable sales, such as size and location. Size and location, as well as financing, conditions of sale, ownership rights, and market conditions (time), are known as elements of comparison. "Elements of comparison are the characteristics of properties and transactions that cause the prices paid for real estate to vary."[8]

Location, of course, has a major effect on value. If at all possible, comparable sales should be selected from within the subject neighborhood or area. Identifying the neighborhood boundaries may be difficult. Sales in the same neighborhood or area as the subject property are affected by the same environmental, economic, social, and governmental forces. However, if necessary, comparables may be selected from different neighborhoods provided the comparables are subject to similar value influences as the subject neighborhood. This concept endorses the use of sales that are located several counties distant from the subject. This practice suggests that comparable sales may be located considerable distances from the subject when the comparables have the same highest and best use and are subject to the same value influences as the subject.

One of the more significant value influences in many areas is the whitetail deer. For example, successful real estate brokers have said that if it were not for the whitetail deer, very few south Texas ranches would be sold, and certainly not at the current prices. Hunting is a common denominator that allows the comparison of ranches in one county with ranches located many miles away in distant counties. Hunting is a major part of the highest and best use of many south Texas ranches and the reason for values well above economic levels. Other game species are also responsible for the current demand for rural properties. The recreational user is addressed more fully in Chapter 3, with Chapters 12 and 13 addressing wildlife issues more fully.

There is no doubt that size, in terms of acres, is a major determinant in the **per acre** value of a property. Typically, a large property will sell for less, on a per-acre basis, than a similar, smaller property. This is due to the supply and demand for rural properties. There are many more buyers in the market for a five hundred acre ranch than for a twenty thousand acre ranch. The number of individuals who can afford five hundred acres is vastly greater than the number who can afford twenty thousand acres. Consequently, the greater number of market participants bid up the price of small properties (on a per-acre basis). The end result is that it is very difficult to compare small properties with large properties in the sales comparison approach. There are no specific guidelines that dictate limits on the size of comparable sales versus the size of the subject property. Common sense as well as the availability of comparable sales must guide the appraising agent.

Other value influences or elements of comparison should be noted in the investigation of the comparable sales. "Market evidence should be tested to identify the variable elements to which property values are especially sensitive."[9] For example, analysis of comparable sales over recent years indicates that properties with large stock tanks or ponds, as the water impoundments are called in some places, typically sell for a greater price than properties without similar water (on a per-acre basis). The presence of a stock tank is an "element of comparison," or something that makes value vary between properties.

Water wells or the availability of groundwater may affect market value. In some areas it is not possible to obtain water from wells due to the absence of groundwater. There are ranches with elaborate residences that have no on-site source of water. Potable water must be purchased by the truckload on a weekly or monthly basis. An appraiser or broker should consider the negative impact on value resulting from the absence of groundwater.

Footnotes

[1]*The Dictionary of Real Estate Appraisal,* Appraisal Institute, Third Edition, 1993, page 67.

[2]*The Appraisal of Real Estate,* Appraisal Institute, Tenth Edition, 1993, page 563.

[3]*The Dictionary of Real Estate Appraisal,* Appraisal Institute, Third Edition, 1993, page 318.

[4]*The Dictionary of Real Estate Appraisal,* Appraisal Institute, Third Edition, 1993, page 403.

[5]*The Dictionary of Real Estate Appraisal,* Appraisal Institute, Third Edition, 1993, page 222.

[6]*The Dictionary of Real Estate Appraisal,* Appraisal Institute, Third Edition, 1993, page 20.

[7]*The Dictionary of Real Estate Appraisal,* Appraisal Institute, Third Edition, 1993, page 207.

[8]*The Dictionary of Real Estate Appraisal,* Appraisal Institute, Third Edition, 1993, page 373.

[9]*The Dictionary of Real Estate Appraisal,* Appraisal Institute, Third Edition, 1993, page 373.

10 Appraising Rural Property—The Analysis

After a representative sample of comparable sales has been researched and verified, it is time to begin the analysis.

Typically, three comparable sales are used in the appraisal. Often, too many sales confuse the analysis and result in a distorted value indication. If an abundance of comparable sales is available, choose the three to four sales that are most similar to the subject property in terms of the elements of comparison that were discussed in Chapter 9.

In the sales comparison approach, a comparative analysis can be used to derive a value indication for the subject property. In the comparative analysis, each of the comparable sales is adjusted for differences in the elements of comparison between the sale and the subject property. Positive adjustments are made for deficiencies in the comparable property relative to the subject property.

Negative adjustments are made for superior characteristics of the comparable property relative to the subject property. In other words, each sale is adjusted for all material differences between itself and the subject property.

A tool that facilitates this analysis is an adjustment grid in which a direct comparison between the subject and sales can be made. On the horizontal axis, the subject property and each of the sales are recorded. On the vertical axis, all of the elements of comparison are listed. An example of a simple adjustment grid appears in Fig. 10-1.

In order to demonstrate a comparative analysis, a hypothetical fact situation has been created in which a subject property is appraised. This example is intended to be an oversimplification to demonstrate basic concepts. The subject and comparable sale data are given in Fig. 10-2. Following the data is an adjustment grid (Fig. 10-3) that provides the comparative analysis.

Figure 10-1

Description	*Subject*	*Sale No. 1*	*Sale No. 2*	*Sale No. 3*
Price per Acre				
Date of Sale				
Adjustment				
Location				
Adjustment				
Land				
Adjustment				
Water				
Adjustment				
Size				
Adjustment				
Access				
Adjustment				
Improvements				
Adjustment				
Total Adjustments				
Indicated Value				

Figure 10-2. Example No. 1.

Description	*Subject*	*Sale No. 1*	*Sale No. 2*	*Sale No. 3*
Price per Acre	No	$700 Cash	$1,000 Cash	$725 Cash
Date of Sale		01/96	06/96	03/96
Location	Average Rural	Inferior Rural	Average Rural	Average Rural
Size	200 Acres	190 Acres	200 Acres	210 Acres
Land	Rangeland/ Native Brush	Rangeland/ Native Brush	Rangeland/ Native Brush	Rangeland Native Brush
Access	FM Highway	FM Highway	FM Highway	Easement
Improvements	Fencing/ Water Well	Fencing	Fencing/ Water Well/ House*	Fencing/ Water Well
Appraisal Date	07/96			

*House has an estimated contributory value of $40,000, or $200 per acre.

Figure 10-3. Example No. 1.

Description	*Subject*	*Sale No. 1*	*Sale No. 2*	*Sale No. 3*
Price per Acre		$700	$1,000	$725
Date of Sale	07/96	01/96	06/96	03/96
Location	Average	Inferior	Average	Average
Adjustment		+$100	$0	$0
Land	Rangeland	Rangeland	Rangeland	Rangeland
Adjustment		$0	$0	$0
Water	Well	None	Well	Well
Adjustment		+$25	$0	$0
Size	200 Acres	190 Acres	200 Acres	210 Acres
Adjustment		$0	$0	$0
Access	Highway	Highway	Highway	Easement
Adjustment		$0	$0	+$50
Improvements	Fencing	Fencing	Fencing/ House	Fencing
Adjustment		$0	–$200	$0
Total Adjustments		+$125	–$200	+$50
Indicated Value		$825/Acre	$800/Acre	$775/Acre

Each of the comparable sales is adjusted to make it more similar to the subject property. Sale No. 1 was adjusted upward because its location is inferior to the subject location and it has no water well. Sale No. 2 was adjusted downward because it has a dwelling and the subject property does not. Sale No. 3 was adjusted upward because it is located on a very long easement through several ranches. After all of the adjustments have been made, the indicated (adjusted) value range (the bottom line) is from $775 per acre to $825 per acre. The estimated value of the subject property must be concluded to be within that range. Most weight should be given to the comparable sale that is most similar to the subject property.

Alternatively, equal weight may be given to two of the sales or to each of the three sales if it is judged that each of the sales is an equally reliable indicator of value. Comparable Sale No. 1 was adjusted approximately eighteen percent ($125/$700), Comparable Sale No. 2 was adjusted twenty percent ($200/$1,000), and Comparable Sale No. 3 was adjusted seven percent ($50/$725). Since Comparable Sale No. 3 was adjusted least, greatest weight should be given to Sale No. 3. This does not mean that Sale No. 1 and Sale No. 2 are to be ignored, but that the value conclusion should be toward the lower end of the value range indicated, say between $775 and $800.

A second example is provided in Fig. 10-4 and Fig. 10-5. The subject data and the comparable sale data are provided in Fig. 10-4. Figure 10-5 shows the adjustment grid which gives the comparative analysis.

Each of the comparable sales is adjusted **to** the subject property for differences in financing, location, improvements, etc. Each sale is adjusted to make it identical (or more similar) to the subject property. Each adjustment is based upon the market reaction to the difference (financing, location, improvement, etc.). If an item in a comparable sale is superior to the subject property, a negative adjustment is made. If an item in a comparable sale is inferior to the subject property, a positive adjustment is made.

Comparable Sale No. 1 was adjusted upward for inferior location and an inferior irrigation well. In order to get an indication of the value of the subject property, Comparable Sale No. 1 required two positive adjustments.

Figure 10-4. Example No. 2.

Description	*Subject*	*Sale No. 1*	*Sale No. 2*	*Sale No. 3*
Price per Acre		$800	$1,025	$900
Date of Sale		06/96	04/96	05/96
Financing		Cash	Cash	Owner Financing
Location	Average Rural	Inferior Rural	Average Rural	Good Rural
Size	300 Acres	300 Acres	275 Acres	325 Acres
Land	Irrigated Cropland/ Good	Irrigated Cropland/ Good	Irrigated Cropland/ Good	Dry Cropland
Access	Paved Highway	Paved Highway	Paved Highway	Paved Highway
Improvements	Irrigation Well 1,000 gpm	Irrigation Well 500 gpm	Irrigation Well 1,000 gpm/ Implement Barn*	None
Appraisal Date	07/96			

Implement Barn has an estimated contributory value of $2,750, or $10 per acre.

Comparable Sale No. 2 was adjusted downward for having an implement barn. In order to make Comparable Sale No. 2 identical to the subject property, the estimated contributory value of the barn was subtracted from the sale price.

Comparable Sale No. 3 was adjusted downward for owner financing and downward for location because both of these factors were considered to be superior relative to the subject property. A positive adjustment was made to Sale No. 3 because it does not have an irrigation well as does the subject property. The adjustment amount of $200 per acre was based upon the market's reaction to the difference between irrigated cropland and dry cropland. The $200 per acre adjustment times the three hundred and twenty-five acres in the sale property equals $65,000. This may, or may not, reflect the cost of an irrigation well in the area. Other factors such as the supply and demand for irrigated cropland may impact the adjustment.

Of the three comparable sales, Sale No. 2 was adjusted least which indicates that it is most similar to the subject. Therefore,

Figure 10-5. Example No. 2.

Description	*Subject*	*Sale No. 1*	*Sale No. 2*	*Sale No. 3*
Price per Acre		$800	$1,025	$900
Date of Sale	06/96	06/96	04/96	05/96
Financing	Cash	Cash	Cash	Owner Financing
Adjustment		$0	$0	–$50
Location	Average	Inferior	Average	Good
Adjustment		+$100	$0	–$75
Land	Cropland/ Good	Cropland/ Good	Cropland/ Good	Cropland/ Good
Adjustment		$0	$0	$0
Size	300 Acres	300 Acres	275 Acres	325 Acres
Adjustment		$0	$0	$0
Improvements	Irrigation Well 1,000 gpm	Irrigation Well 500 gpm	Irrigation Well 1,000 gpm/ Barn	None
Adjustment		+$100	–$10	+$200
Access	Paved Highway	Paved Highway	Paved Highway	Paved Highway
Adjustment		$0	$0	$0
Total Adjustments		+$200	–$10	+$75
Indicated Value		$1,000/Acre	$1,015/Acre	$975/Acre

Comparable Sale No. 2 should be given the greatest consideration. Sale No. 1 was adjusted more than Sale No. 2, but less than Sale No. 3, so it is the next most similar to the subject property. The final value conclusion should be between $1,000 per acre and $1,015 per acre, say $1,010 per acre rounded to $1,000 per acre.

A complete appraisal must consider additional factors such as property rights (mineral rights), financing, condition of sale (arm's-length or distress), and market conditions (time), as well as physical characteristics. In addition, it is also important to consider making adjustments in proper sequence if percentage adjustments are used. An adjustment grid that reflects the proper sequence is shown in Fig. 10-6.

Figure 10-6

Description	*Subject*	*Sale No. 1*	*Sale No. 2*	*Sale No. 3*
Price per Acre				
Sale Price/Acre				
Minerals Adjustment				
Adj. Price/Acre				
Financing Adjustment				
Adj. Price/Acre				
Conditions of Sale Adjustment				
Adj. Price/Acre				
Market Condition/ Time Adjustment				
Adj. Price/Acre				
Location Adjustment				
Size Adjustment				
Water Adjustment				
Land Quality Adjustment				
Road Frontage Adjustment				
Fencing Adjustment				
Structural Improvements Adjustment				
Total Net Adjustments				
Indicated Value				

After it has been determined what adjustments are to be made and the order in which the adjustments will be made, the amount of the adjustments must be considered. Determining how much should be adjusted for size, location, financing, etc. will be explained in the next example.

The subject property for Example No. 3 is a one thousand acre ranch with one-half of the mineral rights, state highway frontage, typical livestock fencing, one water well, and no structural improvements. Also assume that the soil is a sandy loam of average depth and quality and the topography varies from level to gently rolling. The vegetation consists of native brush and the property provides good wildlife habitat. There is no existing mineral production and there are no ingress/egress easements on the property. The highest and best use of the subject is a combination of agricultural use and recreational (hunting) use. The appraisal date is December 1, 1996.

A search of comparable sales was made. The three most similar sales found are presented below. Following the comparable sales is an adjustment grid. Note that the comparable sales are aligned on the horizontal axis and the elements of comparison are aligned on the vertical axis of the adjustment grid.

Sale No. 1 is a five hundred acre ranch located fifty miles southeast of the subject property (in another county) with no mineral rights, easement access, typical livestock fencing, no water source, and no structural improvements. The soil is a clay loam of average depth and quality and the topography varies from level to gently rolling. The vegetation consists of native brush and the property provides good wildlife habitat. There is no existing mineral production and there are no ingress/egress easements on the property. This property sold in June of 1996 for $500 per acre, with owner financing. The sale was an arm's-length transaction. The property has the same highest and best use as the subject property. There is no residential, commercial, or industrial potential.

Sale No. 2 is a one thousand acre ranch located in the same county as the subject with one-half of the mineral rights, state highway frontage, only partially fenced (one mile unfenced), two water wells, and no structural improvements. The soil is sandy loam of average depth and quality and the topography varies from level to gently rolling. The vegetation consists of native brush and the property provides good wildlife habitat. There is no existing mineral production and there are no ingress/egress easements on the property. This property sold in December of 1996 for $550 per acre, cash to seller. The sale was an arm's-length transaction. The property has the same highest and best use as the subject. There is no commercial, residential, or industrial potential.

Sale No. 3 is a fifteen hundred acre ranch located in the same county as the subject property with all of the mineral rights, farm-to-market highway frontage, typical livestock fencing, one water well and one stock tank, and a single-family residence with 1,125 square feet of living area. The soil, topography, and vegetation are the same as the subject. There is no existing mineral production and there are no ingress/egress easements on the property. This property sold in December of 1995 for $700 per acre, cash to seller. The sale was an arm's-length transaction. The highest and best use of the sale property is the same as the highest and best use of the subject. There is no commercial, residential, or industrial potential.

An adjustment grid with the pertinent information is given in Fig. 10-7.

At this point in the analysis, adjustments must be made to each comparable sale for differences between the sale and the subject property. Positive adjustments are made for each element of comparison in which the sale is inferior to the subject. Negative adjustments are made for each element of comparison in which the sale is superior to the subject. The adjustment grid in Fig. 10-7 implies that through analysis of comparable sales, the appraiser concluded that the relevant elements of comparison for the appraisal are as follows: mineral rights (property rights), financing, conditions of sale (arm's-length), market conditions (time), location, size, water, land quality, road frontage, fencing, and structural improvements.

The dollar amount of each adjustment made in the appraisal must be extracted from market data. One of the most common techniques for identifying and estimating the dollar amount of each adjustment is known as "paired sales analysis." In this technique, two ranch sales that are similar in every respect, except one,

Figure 10-7. Example No. 3.

Description	*Subject*	*Sale No. 1*	*Sale No. 2*	*Sale No. 3*
Sale Price/Acre	**None**	**$500**	**$550**	**$700**
Minerals Adjustment	50%	None	50%	100%
Adj. Price/Acre				
Financing Adjustment	Cash	Owner Financed	Cash	Cash
Adj. Price/Acre				
Conditions of Sale Adjustment	Arm's-length	Arm's-length	Arm's-length	Arm's-length
Adj. Price/Acre				
Market Condition/ Time Adjustment	12/96	06/96	12/96	12/95
Adj. Price/Acre				
Location Adjustment	Frio County	Duval County	Frio County	Frio County
Size Adjustment	1,000 Acres	500 Acres	1,000 Acres	1,500 Acres
Water Adjustment	One Well	None	Two Wells	One Well & One Tank
Land Quality Adjustment	Average	Similar	Similar	Similar
Road Frontage Adjustment	State Hwy.	Easement	State Hwy.	F.M. Hwy.
Fencing Adjustment	Average	Average	1 Mile Unfenced	Average
Structural Improvements Adjustment	None	None	None	Residence (1,125 Sq Ft)
Total Net Adjustments				
Indicated Value				

are compared to each other in order to identify the contributory value of the one difference. The contributory value of the difference is used in the adjustment grid to make the sale property more similar to the subject. The two sales used in the paired analysis need not be the same comparables used in the adjustment grid in the primary analysis.

In this hypothetical appraisal, the mineral rights adjustment is analyzed first. Assume the appraiser locates two sales in the subject neighborhood that are similar in every respect except mineral rights. The first sale has one-half of the mineral rights and sold for $550 per acre. The second sale has none of the mineral rights and sold for $500 per acre. Given that the sales are alike in every other respect, the contributory value of one-half of the mineral rights is indicated to be $50 per acre. Other paired sales were then analyzed to verify the conclusion of $50 per acre for the contributory value of one-half of the mineral rights.

The financing adjustment is best extracted from the market by comparing the sale price of the owner financed property (Sale No. 1) to the sale price of a similar property (paired sale) that was transacted on an all cash basis. The difference between the sale prices is due to the effect of the owner financing.

A second method of deriving the financing adjustment involves calculating the cash equivalent sale price. The difference between the cash equivalent price and the owner financed price equals the financing adjustment. For example, Sale No. 1 sold for $500 per acre times five hundred acres for a total sale price of $250,000. The seller carried a promissory note in the amount of $200,000 or eighty percent of the total sale price. The cash down payment was for $50,000 or twenty percent of the total sale price. The interest rate of the note was five percent and the term was ten years. The annual payments are calculated to be $25,901. Assume that the market interest rate at the time of sale was 7.75%. The cash equivalent price is calculated by adding the present value of the annual payments discounted at 7.75% to the cash down payment of $50,000.

The present value of ten annual payments of $25,901 discounted at 7.75% is $175,775. The cash down payment of $50,000 is added to present value of the note in the amount of $175,775 giving a cash equivalent price of $225,775. The cash equivalent price is subtracted from the owner financed price of $250,000 giving an adjustment of $24,225 or $48.45 per acre ($24,225 divided by 500 acres = $48.45) rounded to $50 per acre.

This method of deriving the financing adjustment gives the appearance of being mathematically precise and therefore absolutely correct. This is not necessarily the case. There may be underlying buyer or seller motivations for owner financing. Therefore, the best method of determining the financing adjustment is usually by paired sales analysis.

Before the adjustment process begins, the proper sequence of adjustments should be addressed. Current appraisal theory holds that the sequence of adjustments should be determined by the **market.** The sequence may be difficult to determine. A recommended sequence of adjustments is as follows:

1. Property rights (mineral rights)
2. Financing
3. Conditions of sale (arm's-length versus distress)
4. Market conditions (time)
5. All others, any order

Current theory suggests that when percentage adjustments are made, the first four adjustments listed above should be made individually and in the order given. The adjustment for mineral rights should be made to the sale price prior to the financing adjustment. The financing adjustment should be applied to the product of the previous adjustment and so on. After the adjustment for market conditions (time) is made, the remaining adjustments may be made on a net basis. In other words, the adjustments for location, size, water, land quality, road frontage, fencing, and structural improvements are summed and applied to the product of the first four adjustments.

An adjustment grid showing the results of the adjustment process is given in Fig. 10-8.

A discussion of the adjustments is given below. Each of the adjustments has been derived through paired sales analysis. The ad-

Figure 10-8

Description	*Subject*	*Sale No. 1*	*Sale No. 2*	*Sale No. 3*
Sale Price/Acre	**None**	**$500**	**$550**	**$700**
Minerals	50%	None	50%	100%
Adjustment		+$50	$0	–$50
Adj. Price/Acre		**$550**	**$550**	**$650**
Financing	Cash	Owner Financed	Cash	Cash
Adjustment		–$50	$0	$0
Adj. Price/Acre		**$500**	**$550**	**$650**
Conditions of Sale	Arm's-length	Arm's-length	Arm's-length	Arm's-length
Adjustment		$0	$0	$0
Adj. Price/Acre		**$500**	**$550**	**$650**
Market Condition/ Time	12/96	06/96	12/96	12/95
Adjustment		–$30	$0	–$78
Adj. Price/Acre		**$470**	**$550**	**$572**
Location	Frio County	Duval County	Frio County	Frio County
Adjustment		+$25	$0	$0
Size	1,000 Acres	500 Acres	1,000 Acres	1,500 Acres
Adjustment		–$47	$0	+$29
Water	One Well	None	Two Wells	One Well & One Tank
Adjustment		+$5	–$5	–$5
Land Quality	Average	Similar	Similar	Similar
Adjustment		$0	$0	$0
Road Frontage	State Hwy.	Easement	State Hwy.	F.M. Hwy.
Adjustment		+$47	$0	$0
Fencing	Average	Average	1 Mile Unfenced	Average
Adjustment		$0	+$5	$0
Structural Improvements	None	None	None	Residence (1,125 Sq Ft)
Adjustment		$0	$0	–$15
Total Net Adjustments		+$30	$0	+$9
Indicated Value		**$500**	**$550**	**$581**

justment process endeavors to make each sale property like the subject property. Positive adjustments are made for deficiencies in the sale property relative to the subject property. Negative adjustments are made for superior characteristics of the comparable sale relative to the subject. The derivation of the mineral rights adjustment and the financing adjustment were demonstrated in the preceding pages.

MINERAL RIGHTS

The subject property has fifty percent, or one-half, of the mineral rights. Sale No. 1 is located in a similar mineral area and has no mineral rights, so it was adjusted plus $50 per acre. Sale No. 2 is located in the subject neighborhood and has one-half of the mineral rights (as does the subject property), so it was not adjusted. Sale No. 3 is located in the subject neighborhood and has one hundred percent of the mineral rights, so it was adjusted minus $50 per acre.

The adjusted prices of Sales No. 1, 2, and 3 are $550 per acre, $550 per acre, and $650 per acre, respectively.

FINANCING

The subject property is appraised on a cash to seller basis. Sale No. 1 was owner financed. Through investigation of the terms of the note and analysis of comparable sales it was determined that the owner financing was advantageous to the purchaser of the property and that the sale price reflected a premium for the owner financing. Analysis of paired sales indicated that a $50 per acre premium was paid for the owner financing. Therefore, Sale No. 1 was adjusted minus $50 for financing. Sale No. 2 and Sale No. 3 were on a cash basis and so were not adjusted.

The adjusted prices per acre of Sales No. 1, 2, and 3 are $500 per acre, $550 per acre, and $650 per acre, respectively.

CONDITIONS OF SALE

Each of the comparable sales was found to be on an arm's-length basis so no adjustments were required.

MARKET CONDITIONS/TIME

Market data indicated that property values in the subject area are decreasing at a rate of one percent per month (remember, this is a hypothetical example!). Sale No. 1 occurred six months prior to the appraisal date and so was adjusted minus six percent times the adjusted price per acre of $500, giving a minus adjustment of $30 per acre. Sale No. 2 occurred in the same month as the date of the appraisal of the subject property and so was not adjusted. Sale No. 3 occurred twelve months prior to the date of appraisal, so it was adjusted minus twelve percent times the adjusted price per acre of $650, giving an adjustment of minus $78 per acre.

The adjusted price per acre of Sales No. 1, 2, and 3 are $470 per acre, $550 per acre, and $572 per acre, respectively.

The remaining adjustments are to be summed and applied to the adjusted sale prices on a net basis.

LOCATION

The subject property and Sales No. 2 and 3 are located in the same general area of Frio County. Therefore, Sales No. 2 and 3 do not require a location adjustment. Sale No. 1 is located in the northern portion of Duval County. Paired sales analysis indicated that the location of the subject property is more valuable relative to Sale No. 1 in the amount of $25 per acre. Therefore, Sale No. 1 was adjusted plus $25 per acre.

SIZE

The size adjustment is difficult to estimate and more difficult to explain. It is a certainty, however, that smaller properties typically sell for a greater price per acre than do larger properties. This is because there is a greater number of market participants seeking to purchase small ranches relative to large ranches. The greater number of buyers tend to compete and bid prices up. It is necessary to briefly diverge from the appraisal to explain a technique commonly used to estimate the size adjustment.

Many south Texas appraisers attempt to measure the applicable size adjustment based upon a technique involving "doubling in size." The theory states that for each doubling in size (times two) of a ranch there is a corresponding decrease in price per acre. Typically, the adjustment is in the range of five percent to ten percent per doubling, times the value or price per acre of the comparable sale.

For example, assume a five thousand acre ranch sold for $300 per acre. Also assume the adjoining ranch is being appraised, which is ten thousand acres in size. The two ranches are alike in every respect except size. The subject property is twice (times two) as large as the comparable sale or an amount equal to "one doubling." (For our example, the size adjustment will be ten percent times the comparable sale price for each doubling in size.) The adjustment amount is minus ten percent times $300 per acre or minus $30 per acre. The sale property is smaller than the subject so the adjustment is "minus."

If the subject contained twenty thousand acres, the size difference would be "two doublings" (5,000 acres × 2 = 10,000 acres × 2 = 20,000 acres), therefore the adjustment would be minus twenty percent (10% × 2) times $300 per acre or minus $60 per acre.

Assume a ten thousand acre subject is compared to a twenty thousand acre comparable ranch sale which sold for $245 per acre. The size difference between subject and comparable is one full doubling. The size adjustment is calculated to be $245 times ten percent or plus $24.50 per acre. The sale property is larger than the subject so the adjustment is "plus."

Now return to the example appraisal. Assume paired sales data indicated the appropriate size adjustment to be ten percent for each doubling in size. The size difference between the subject and Sale No. 1 is one full doubling (one thousand acres versus five hundred acres). Therefore, the size adjustment to Sale No. 1 is minus ten percent times the adjusted sale price of $470 per acre or minus $47 per acre. The subject is larger than the comparable so the adjustment is "minus." There is no size difference between the subject and Sale No. 2, so no adjustment was made.

The size difference between the subject and Sale No. 3 is one-half of one doubling (one thousand acres versus fifteen hundred acres) so the adjustment is plus five percent (1/2 × 10%) times $572 per acre, or plus $29 per acre. The subject is smaller than the comparable sale so the adjustment is "plus."

WATER

The subject property has one water well. Analysis of sales indicated that the contributory value of the one water well was approximately $5,000, or $5 per acre for the one thousand acres contained in the subject property. Sale No. 1 has no water wells and no source of water so was adjusted plus $5 per acre. Sale No. 2 has two water wells which is superior to the subject property by one water well so was adjusted minus $5 per acre. Sale No. 3 has one water well and one small stock tank. The stock tank was estimated to have approximately the same contributory value as one water well, so Sale No. 3 was adjusted minus $5 per acre.

LAND QUALITY

The subject property was considered to have average quality soil and topography. Each of the sales is considered to have soil capability sufficiently similar to the subject property so that no adjustments were required.

ROAD FRONTAGE

The subject property is located on a state highway. An analysis of comparable sales indicated that properties located on state and farm-to-market highways sold for a greater amount on a per-acre basis as compared to off-road properties with easement access. Analysis indicated a difference in sale price of ten percent. Sale No. 1 is located on an unpaved easement, so it was adjusted ten percent times $470 or plus $47 per acre.

FENCING

The subject property has typical livestock fencing in average condition. Sale No. 1 and Sale No. 3 had equal fencing as of the sale date so no adjustments were required. Sale No. 2 had one mile of its perimeter unfenced. The contributory value of the fencing was found to be approximately $5,000 or $5 per acre. Therefore, Sale No. 2 was adjusted plus $5 per acre.

STRUCTURAL IMPROVEMENTS

The subject property has no structural improvements. Sale No. 1 and Sale No. 2 had no structural improvements and so were not adjusted. Sale No. 3 had a modest ranch residence with a gross living area of 1,125 square feet. Sales of similar properties with similar ranch residences were analyzed. The residences were found to have contributory values of approximately $20 per square foot of gross living area. The gross living area contained in the residence of Sale No. 3 in the amount of 1,125 square feet was multiplied by the $20 per square foot, giving a total contributory value of $22,500. This amount was divided by the fifteen hundred acres in the sale property, giving a minus adjustment of $15 per acre.

The indicated values of Sale No. 1, 2, and 3 are $500 per acre, $550 per acre, and $581 per acre, respectively.

RECONCILIATION

At this point in the appraisal, the agent must conclude a value from the adjusted sale prices. The agent should review the comparable sales and then give the greatest weight to the sale or sales that are most similar to the subject property. This will be the sale or sales that have required the least amount of adjustments. As seen on the adjustment grid, Sale No. 2 is the most similar to the subject property in terms of mineral rights, date of sale, location, size, etc. This is evident since Sale No. 2 required the least adjustments on the adjustment grid. Therefore, it may be concluded that Sale No. 2 is the best indicator of value for the subject property.

Based upon the sales analysis, the market value estimate of the subject property is $550 per acre for the one thousand acres contained in the subject property or a total market value estimate of $550,000.

It should be remembered that real estate appraisal is not an exact science and that the quality of the comparable sales used in the analysis determines the accuracy and quality of the appraisal. The majority of the real estate practitioners' time and effort should be spent researching comparable sales. The agent who is the most knowledgeable about comparable sales in the neighborhood will produce the best estimate of market value.

The income approach to value involves capitalizing the estimated net operating income of the subject property. The formula for the income approach is shown in Fig. 10-9.

Gross income may be estimated based upon cash rentals or from production budgets. The object is to estimate the income attributable to the land. Expenses include real estate taxes, management, maintenance, and reserve for repairs (fence, roads, pens, water wells, etc.). Capitalization rates are extracted from comparable sales and calculated as follows: Net Operating Income divided by Sale Price equals Capitalization Rate.

An example of the income approach is illustrated next. The subject property is a twenty-five hundred acre ranch consisting entirely of rangeland. Income may be derived from hunting and grazing leases. The estimated economic income from hunting is \$7 per acre per year. The estimated grazing income is \$3 per acre per year. Expenses are estimated to be approximately twenty percent of gross income, or \$2 per acre per year. The net operating income is therefore \$10 minus \$2 equals \$8 per acre per year.

These comparable sales were researched to calculate capitalization rates. The calculations are shown in Fig. 10-10.

Figure 10-9

Gross Income – Expenses = **Net Operating Income**

Net Operating Income ÷ Capitalization Rate = **Market Value**

1.
Less: Gross Income
<u>Expenses</u> (Taxes, Maintenance, Management)

= **Net Operating Income**

2.
Divided By: Net Operating Income
<u>Capitalization Rate</u>

= **Market Value**

Figure 10-10

Sale Number	*Size*	*N.O.I./Acre*	*Price/Acre*	*Cap. Rate*
1	3,000 Acres	$7	$400	1.8%
2	2,500 Acres	$8	$425	1.9%
3	2,000 Acres	$9	$450	2.0%

N.O.I. = Net Operating Income
Cap. Rate = Capitalization Rate

A capitalization rate of 1.9% was selected because Sale No. 2 is most similar to the subject property in terms of physical characteristics. The calculation of the value estimate of the subject property is shown in Fig. 10-11.

A problem with the income approach for range properties in some areas is that the capitalization rates are so low that a minor change in the rate results in a large change in value. Also, if buyers and sellers are not using the income approach to estimate value, it could be misleading for appraisers or brokers to attempt to do so.

Estimating the net income attributable to land from a production standpoint should theoretically yield the same results as using cash rents. Using the same property for illustration, assume a cow/calf operation with a stocking rate of twenty-five acres per animal unit. At this rate, the subject property should be able to support one hundred animal units.

For illustrative purposes, assume that the subject ranching operation will produce 90 calves of which 45 will be steers and 45 will be heifers. Ten replacement heifers are retained and ten cull cows are sold. The estimated market value for the ranch may be calculated as in Figure 10-12.

Figure 10-11

Economic Gross Income	Operating Expenses	Net Operating Income	Cap. Rate	Market Value
$10	– $2	= $8	÷ 0.019	= $421

Figure 10-12

Steers—	45 × 500 lbs. × $0.65 =	$14,625
Heifers—	35 × 475 lbs. × $0.55 =	9,144
Cows—	10 × 800 lbs. × $0.30 =	2,400
Hunting Income 2,500 Acres × $7 per Acre =		17,500
		$43,669
LESS EXPENSES		
Operating Expenses (30% of Gross)		13,100
Livestock Cost (Amortized Cost)		7,500
Management—5%		2,000
Net Operating Income		$21,069 or $8.43 /Acre

Value Indication by the Income Approach:

Net Operating Income	÷	Capitalization Rate	=	Market Value
$8.43	÷	0.019	=	$444 per Acre

Estimating net operating income using production data is difficult due to the large number of variables involved in the process. The $444 per acre value estimate is considered to be less reliable than the estimate produced by the cash rental method. Both methods are typically less reliable than the sales comparison approach for properties in south Texas. Livestock prices vary from month to month and operating costs change dramatically depending on stocking rates, intensity of management, feed costs, etc. If the highest and best use of the land is solely agricultural, more productive land should be more valuable than less productive land. Better soils and vegetation translate into higher stocking rates and greater net income, therefore, higher prices. However, when the highest and best use of the land is largely recreational (hunting and fishing), soil quality and vegetation are important factors to the extent these factors affect antler size and quality. Cattle stocking rates may have little affect on the market value of rangeland selling at capitalization rates of two percent or less.

IV

UNDERSTANDING THE LAND AND ITS PROPER MANAGEMENT

11 Ecological Aspects of Rural Real Estate

There is a saying among old ranch agents that when out of pocket, they are "out selling dirt." In no other segment of the real estate industry is this more true. Rural real estate involves selling dirt, grass, bushes, wild animals, as well as future production of the land. It involves truly selling the land. Commercial real estate sells tracts for building, storing, or manufacturing. Residential real estate sells tracts with homes or for the placement of new dwellings. Rural real estate is selling a living, dying, growing portion of the Earth for the purpose of producing food, fiber, or aesthetic enjoyment.

In this chapter we will consider those ecological factors that should be addressed when "selling dirt" and how you can more effectively market rural properties by becoming knowledgeable about what the buyer can expect from a purchase. We will also look at differing methodology for the management of rural land to increase its productivity, beauty, or value.

"Selling dirt"

Ecological areas are portions of the earth's surface that are similar in weather, soils, and topography. Such similarities usually result in similar vegetation, if not by species, at least by vegetative type (e.g., grassland or jungle). An example of a map of ecological areas is given in Fig. 11-1.

As a real estate professional specializing in rural properties, the practitioner will be working in one or more ecological areas. By learning the components that contribute to the distinction of each area, practitioners can better predict capabilities and characteristics of a given parcel.

LAND QUALITIES AFFECTING UTILIZATION

Location

A major consideration when buying or selling any real estate is the location. When buying rural real estate, this factor is even more critical as location often dictates the uses of rural land. For exam-

Figure 11-1

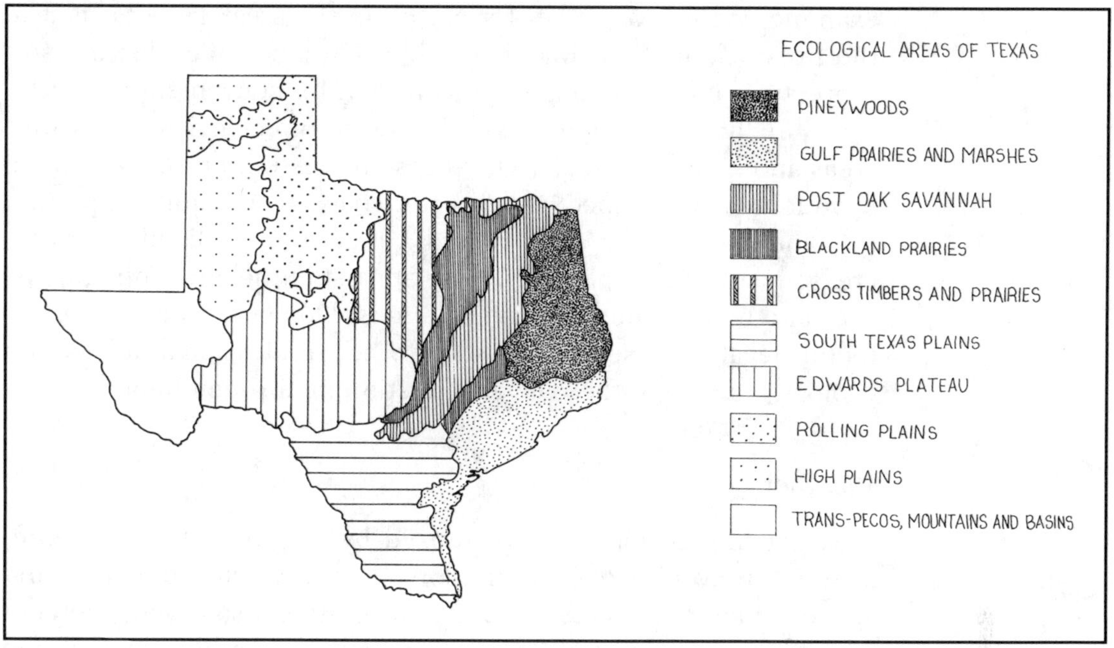

ple, farmers rarely look to expand into Death Valley and ranchers generally shy away from Manhattan; either the land is unproductive or the going price precludes their making a profit. It is a fact that the Rockefellers run cattle on their estate in upper New York; good cattle on productive soil. However, if they had to do that for a living, it would be hard to do on land worth $100 per square foot!

As a general rule, buyers will determine the area where they want to purchase land. From experience, research, or hearsay, buyers often know area what best suits their needs. Do not try to steer buyers away from an area where they want to make a purchase unless you really know your stuff. It has often been the case that buyers who seemed ill-informed in fact not only were correct, but knew the area better than the practitioner. If a practitioner were to attempt to push them toward another property, that would damage the buyers' faith in the practitioner. This does not relieve practitioners of their responsibility to know their product, however. Practitioners who want to sell agricultural land must have at least a basic knowledge of the industry.

Another consideration in regard to location is the market. For example, is there a market for cotton in the area? How far is it to the livestock auction ring? If the buyer wants to develop a resort property, is it close enough to civilization for convenient access?

Finally, regarding location, be aware of ecologically sensitive areas and be informed as to the uses and restrictions on such land in your area. This aspect of real estate has become quite important in some parts of the country and a practitioner's liability has increased with this importance. Stay abreast of the environmental issues and be informed about regulations. Do not hesitate to recommend legal counsel or to seek other professional advice on ecological issues if there is any question in anyone's mind.

Weather

Weather in a region is characterized by long-term records representing an average of precipitation, humidity, wind direction and speed, temperatures, etc. Weather also affects soil types, percent ground cover, type of ground cover, and many other physical and vegetative factors. It is a truism in South Texas that the ten-year annual rainfall average as depicted in Fig. 11-2 includes nine years of drought followed by a hurricane.

While it is true that people cannot change the weather, it is important for practitioners to be knowledgeable and factual in representing it to the buyer. In addition to annual precipitation, the practitioner should be conversant in seasonal changes, highs and lows to be expected, and hazards that might occur. This is particularly important if weather will affect the planned uses of the land.

Soils

Dirt, or soil, is made up of living and dead organisms mixed with inorganic materials derived from parent material or rock. The type of soil depends on the parent material, climate, organisms past and present, location, and time. Soils, as depicted in Fig. 11-3, are classified as to texture by the percentage of clay, loam, sand, silt, gravel, or rocks in the soil and composite names identify these per-

Figure 11-2

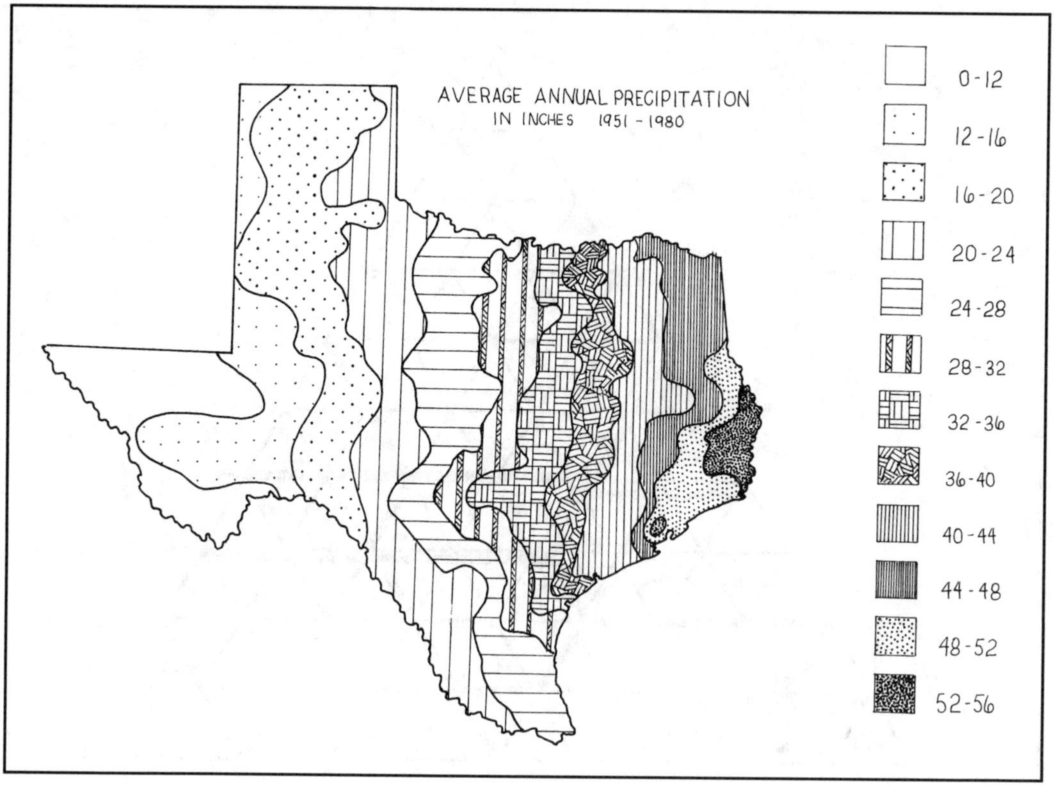

centages. For example, the term sandy clay loam identifies the relative percentages of each of the three soil components. These characteristics help the buyer estimate the productivity of the soil and the degree to which climate will help or hurt the buyers' plans for the property.

Other characteristics of soil important to a buyer include erodibility, either by wind or rain, permeability, the rate of absorption of precipitation, salinity, pH, and available water capacity, i.e., how much water a soil type can hold and release for use by plants.

Water

The most fertile soil in the world will not grow crabgrass without water. Indeed, much of the Earths' surface is not productive specifically because of moisture limitations. When selling rural real estate,

Figure 11-3

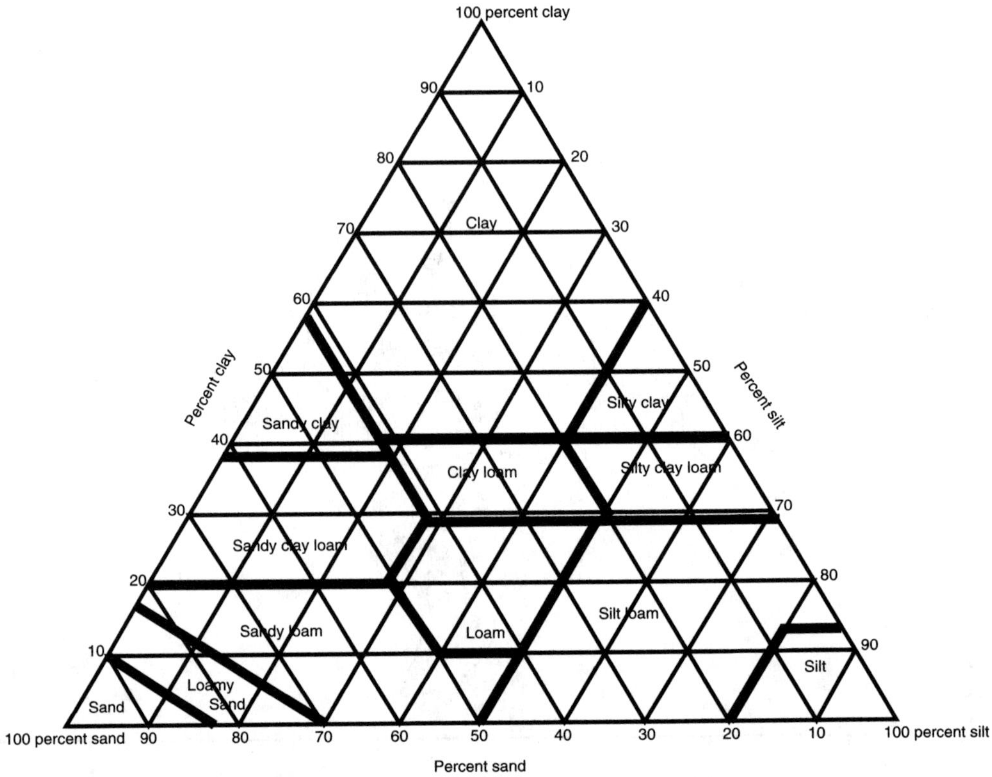

the practitioner must become familiar with the water situation and study ways to improve limited water resources. Even in those cases when the purchase is purely aesthetic, water plays a pivotal role.

Water is available in two forms, surface and subsurface. Surface water exists in lakes, steams, and reservoirs, while well water comes from aquifers or underground bodies of water, as depicted in Fig. 11-4. Numerically, most of the people in the United States get their drinking water from surface reservoirs or lakes. A reservoir is a water impoundment specifically built to hold and reserve water. Conversely, most rural inhabitants get their drinking water from wells. Livestock fall into both categories, although recent developments in efforts to reduce the introduction of effluents into surface water may result in limiting livestock to trough water in the near future.

Figure 11-4

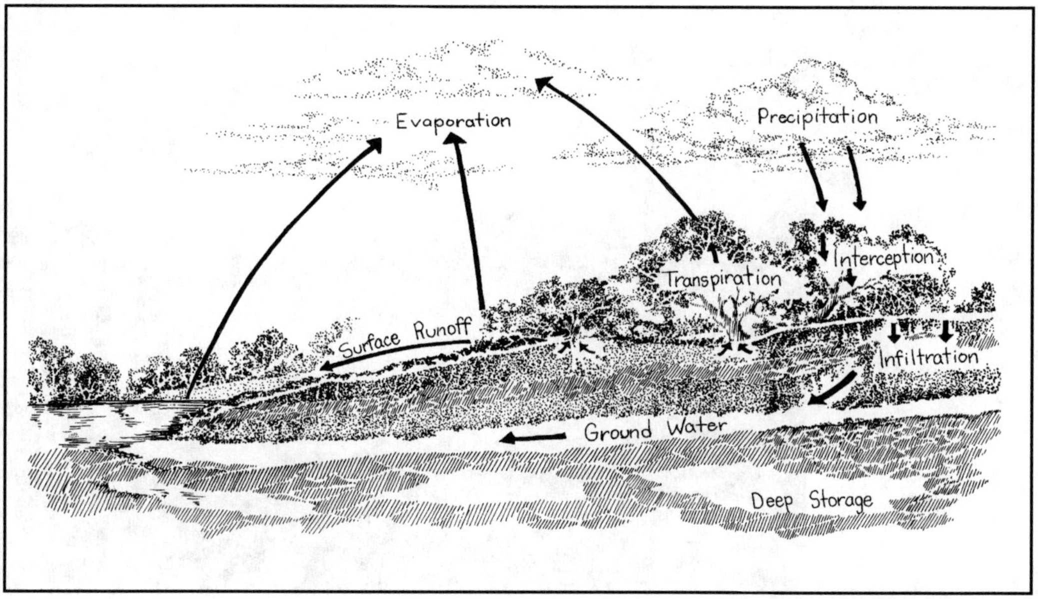

Surface water is exposed to the elements and is therefore highly variable. The practitioner needs to know what the depth of the surface water is, the annual rainfall, the size of the watershed or area from which the water is collected, and the annual loss to evaporation. Knowing this, plus the quality of the streambed or lake bed, the practitioner will be able to determine the dependability of the surface water. For example, a rough estimate of watershed indicates that it takes 100 acres with 25 inches of precipitation to supply one acre-foot of water in an impoundment after deductions for percolation and evaporation. A water-tight pond with a surface area of one acre that averages ten feet in depth equals ten acre-feet. Knowing this, the practitioner realizes that to fill and maintain a ten acre-foot pond, one thousand acres of watershed and twenty-five inches of precipitation are required.

Subsurface water is another story. While the depth and thickness of an aquifer are easily determined by drilling, the potential water volume is not. Production is determined by the size of the gravel or porous rock through which the water is being drawn, depth to the water, and size of the bore. Water wells can be equipped with windmills, solar electric pumps, electric pumps, or

Ponds can be a source of water as well as recreation

combustion engine pumps. Ordinarily, forty gallons per minute are adequate for domestic use while irrigated farming may require up to 2,000 gallons per minute. Water quality is of utmost importance and water should always be tested by qualified personnel for potability and dissolved minerals.

Topography

One of the more important factors in rural real estate that is often overlooked is topography. This refers to the changes in elevation on a tract of land. Topography ranges from the billiard-table-flat farmland of the Midwest to the vertical impassibility of the Rocky Mountains. Topography can make a small place seem bigger by limiting the line of sight, as well as shrink a bigger place whose four corners can all be seen from the bed of a truck.

Topography plays an important role in development of a parcel as well. Row crop farming is, of course, greatly affected by the slope of the land, and erosion is always a consideration when clearing or farming steeper sites. An example is a property that was recently shown that was 9,000 acres of which 650 acres were "usable," the broker's term for trafficable in a normal vehicle. Imagine not being able to understand a topographical map, and showing such a tract to a farm corporation!

A topographical map or "topo" is a wealth of information and is considered a requirement in every sales package. Likewise, the knowledgeable practitioner must understand a topo map and what information it contains. For example, elevation is depicted using contour lines at set elevations as shown in Fig. 11-5. If the elevation lines are close together, it means there is a cliff or steep rise or drop. Conversely, lines far apart indicate a gentle slope. Lines in concentric circles indicate that there is a hill or depression while a long oblong of concentric lines indicates a ridge. Topo maps also show roads, buildings, water, vegetation, and fences. Topo maps are valuable tools to the rural real estate professional, just as vital as a calculator.

Figure 11-5

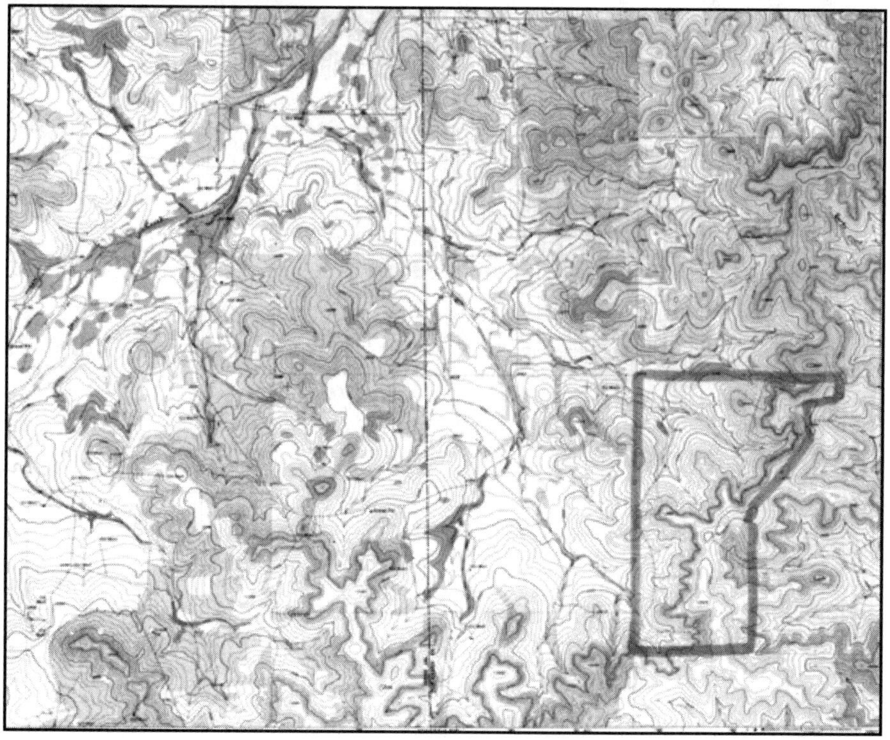

Vegetation

Under normal conditions, vegetation in a given soil type with a given climate will strive to attain a plant community that most efficiently utilizes the potential of the site. Starting with weeds in most cases, the plant community changes with time, gradually reaching the optimum or climax stage. This process is called succession and the various levels are called successional stages with weeds being the first successional stage. As noted, the climax stage is determined by all those factors that impinge on plant life. In a prairie, grasses are the optimum population; on the east coast, hardwood forests would be the climax plants and in the Sonoran desert, the best plant community is drought- and heat-tolerant vegetation such as cacti.

Knowing the area in which a practitioner is working should allow the practitioner to tell at a glance which successional stage the vegetation is in. This is important to the buyer in that the successional stage of the plant community will determine the type of vegetation and its productivity for livestock and wildlife. Cattle ranchers, for example, prefer grass as the climax vegetative state and will reverse succession to maintain grasslands if that is not the normal climax population. Bobwhite quail managers, on the other hand, want that first successional stage as bobwhite quail are creatures of early succession.

Changing the plant community is difficult at times, easy at others. To return to the first stage, simply conduct a hot burn through the community when conditions are right. If there is not enough fuel for a fire, spray the vegetation with herbicide, let it die back, then burn. In the case of forests, clear-cutting, while not politically correct, will result in a plant community of weeds and early-growth woody plants that is ideal for most wildlife.

Mechanical means can also be used to reverse or retard the successional status of a plant community. Simply pushing all the vegetation into piles, called stacking, will promote grass and weeds and can be done selectively to protect larger plants such as trees or certain vegetative types such as a flower species. Other mechanical treatments that simply remove the portions of the plants that

Helicopter applying a brush herbicide

Grass explosion after brush was sprayed

are above ground include shredding, chaining, discing, and roller-chopping. All these techniques promote early succession plantlife by reducing loss of moisture to the larger plants, reducing shading, and stimulating seed germination. Top-removal techniques will result in less of some plants in the future, while other plants have developed defense mechanisms that result in a stronger, larger plant several years post-treatment.

Plant succession can also be affected by managing the intensity and duration of grazing. Heavy grazing will return the community to an early successional stage while prudent grazing can actually promote climax by favoring climax species.

UTILIZATION CONSIDERATIONS

The Real World

A call comes in from potential buyers early Thursday morning. The buyers want a weekend farm with the potential for fruit trees and some cows in addition to irrigated fields. What is the first thing to know in order to present the buyer with a valid selection? First of all, you need to know or find out if fruit trees can be grown in the area the buyer prefers. Is the weather suitable for trees or farming or even cattle? Are there similar operations in the area already? If not, you will need to try to interest the buyers in an area where they will grow, or find out if fruit trees are really all that important to the buyers' plans.

Next, look through your property files for land that is relatively flat but has good drainage, as cattle do not do well on uneven terrain. Bottomland would be good for both cows and farming, unless it is subject to extensive and long-term flooding. In the case of such flooding the land would be unavailable for grazing and crops would be lost. The soil should be a sandy loam for most crops, free of stone and, again, fairly level if the buyers plan to farm. It need not be in fields, as clearing is not cost-prohibitive in certain cases, all else being favorable. Water is very important to these buyers, so you must find properties with water of good quality and quantity, either surface or subsurface. The water must be dependable, par-

Tractor "armored" for shredding brush

Thick brush that is shredded grows quickly

ticularly during periods of high demand, and it must be cheaply accessible as excessive costs in lifting or moving water can devastate the bottom line.

Equipment is not important and may be a detriment if it is in place but not in good repair or not of modern design. Last, you need to consider market options. Are there other producers in the area? Is there someplace for the buyers to sell their produce? What good is raising apples at $2 per bushel if it takes another $2 to get the apples to market and the competition is selling apples at $3 per bushel?

Suppose that, after reviewing the available properties, the buyers decide that they are not cut out for all of the work involved in farming and then decide instead to make money with livestock. In addition to the above factors, the practitioner should determine what type of livestock the buyer plans to run: cows, sheep, zebras, etc. Also, what class of that type: young only, mother and young pairs, males only, females only? Will the buyers have only cows or cows and sheep? Will weather affect their herd so much so that heavy vegetation is required for cover? Will the buyers be "yarding" the animals, putting the livestock in pens and feeding them? And if so, will the runoff from the pens affect water quality?

Livestock

One of the strongest incentives in the purchase of real estate is pride of ownership. This is perhaps even stronger in rural real estate. The buyers want to control, manage, and see results from their efforts. Easily the greatest use of rural real estate in the United States is livestock husbandry where management directly dictates results. From the lush grasslands to the scrub brush of the desert, there is a potential to produce income from livestock.

Livestock is categorized by type and class as well as species. Type refers to the animal form—cattle, sheep, goats, fish, etc.—while classes refer to what growth stage or management protocol are used. Last, the particular species preferred further describes the operation. For example, landowners may choose to have or "run" cattle on their property; that is the type of livestock. The landowners further decide to manage a cow/calf operation, producing calves for sale at weaning. This is the class of livestock. Last, the landowners choose Brangus, a species, as the preferred cow. Likewise, a farmer in Mississippi chooses to raise fish (type), for sale as fingerlings (class), and prefers catfish because of its hardiness (species).

Certainly, there are combinations of the above, as most farm/ranch operators realize that diversity is an advantage. As real estate professionals, practitioners need to be able to discuss the types and classes of livestock suitable to a piece of ground and learn the requirements and limitations of each.

First and foremost, it must be determined how much of the animals' diet will be produced by the land. The simplest example of this would be cows. Usually, cows are allowed to graze all year with supplementation of mineral and energy during periods of harsh weather. In certain parts of the country, such as North Dakota, livestock must be removed or sheltered during the harshest time of winter, and require full dietary supplementation at such times. It behooves practitioners to learn when and how much supplementation is the norm in their area.

Remember, too, that all land is not alike. Livestock managers are simply farmers in that the production of the native or introduced forage plants will determine the numbers of animals they can run in a given year. This is a combination of two factors, the carrying capacity of the land and the management of grazing or consumption of the land's production. Carrying capacity is a fluid concept that changes with succession, weather, and use. By definition, it means how many "consumers" can live on a given piece of land for a given period of time, subsisting on the vegetation. It is usually expressed in Animal Units (AU) over time. For example, one cow and its calf are one animal unit or AU. What the animals require in terms of acreage of forage to do well for one month is one AU per month, or one AUM. Likewise, what that pair of animals requires for one year is written as one AUY. Examples of animal units are given in Fig. 11-6.

Too many cows grazing for too long destroys the land

How does this translate into how many individual animals a tract can support? As an example, the carrying capacity for cattle varies from one cow (1 AUY) to 2 acres in good grasslands, to 1 AUY per one section (640 acres) in the arid Southwest. Since cows are primarily grass eaters, the percent of grass as opposed to brush, trees, or bare ground on a property directly influences the carrying capacity. Last, competition between livestock and wildlife must always be considered, particularly when wildlife is tapped as a source of income.

The introduction of exotic animals is a relatively new area of livestock management with new and unresolved aspects. Most hoofed exotics have much more varied diets than native wildlife or livestock and thus pose particular problems with regard to competition for food. The competition for food between wildlife and domestic livestock occurs only when overgrazing forces live-

Figure 11-6

CATTLE	
Weaned calves to yearlings	0.6 Animal Units
Steers and heifers (1 to 2 years)	1.0 Animal Units
Mature cows with or without unweaned calves	1.0 Animal Units
Bulls (2 years and over)	1.3 Animal Units
SHEEP	
5 weaned lambs to yearlings	0.6 Animal Units
5 mutton or ewes (1 to 2 years)	1.0 Animal Units
5 mature ewes with or without unweaned lambs	1.0 Animal Units
5 rams	1.3 Animal Units
GOATS AND DEER	
6 weaned kids to yearlings	0.6 Animal Units
6 mutton or does (1 to 2 years)	1.0 Animal Units
6 does with or without kids	1.0 Animal Units
6 bucks or mutton over 2 years	1.3 Animal Units
6 whitetail deer	1.0 Animal Units

stock to eat brush and weeds. Then the livestock comes into direct competition with wildlife. Many exotics are omnivorous herbivores, however, eating a wide range of plant life, depending on palatability. When weeds are young and succulent, the exotics eat weeds, thereby competing with wildlife. When grass is actively growing and of optimum value for cattle, the exotics eat grass. Finally, when things get tough and food becomes limited, the exotics eat grass, weeds, brush, or anything, thereby giving the exotics an advantage over both livestock and native wildlife. Caution should be exercised when presented with an exotic species, either in place on a tract of land or as part of a proposed management plan by a buyer. Carrying capacity and densities of exotic livestock/wildlife are often not known and stocking rates are often set too high, to the detriment of native wildlife, livestock, and the range.

Buffalo on a ranch in South Texas

Soils play a major role in considerations of stocking rates and management. Heavier soils are less prone to drought, while sandier soils can respond to less moisture, but do not hold it near the root zone. Nutritive quality of forage produced is a factor as well. Normally, in soils that produce high quantities of forage, as in wetter areas, the forage is less nutritious than in drier areas due to leaching of nutrients. Often, livestock programs in coastal areas require extensive supplementation of minerals and protein. Drainage is important to an animal that must stay "outside" all year. Bottomlands, while generally more productive, may be prone to flooding and are lost to the landowner when inundated. Likewise, vertical slopes are of less value to a cattleman, while perfectly suitable to a sheep manager.

Water sources should be of good quality and dependable. Most livestock managers prefer well water if it can be brought to the surface cheaply. Rivers present a danger of drowning and are difficult to fence while stock tanks or ponds can run dry just when needed most. In most parts of the country, an allowance of 16 gallons per day per AU of good, clean water is acceptable.

The physical plant in place is important to anyone considering a parcel of land for livestock potential. The existence of fences, watering facilities, and working or containment structures (pens) can make or break a deal. As a professional, the practitioner should be aware of the cost of installing the different types of fencing, water wells and troughs, piping, and working pens in the area. Ordinarily, a buyer who plans to implement either fish farming, (aquaculture) or contained livestock (hogs, chickens, etc.) will be aware of the costs of required structures. If such structures are in place on a tract, leave it to professional appraisers to determine their value.

Last, the practitioner should be aware of the location of markets for various livestock types in the area. Certainly, the practitioner can never give estimates of potential income, but the practitioner should be able to describe marketing businesses and provide the names and numbers of principal marketing professionals as well as livestock transporters. If something can't be sold, it has no value. This applies to livestock as well as real estate.

Wildlife

In conjunction with its aesthetic beauty, perhaps the greatest value of rural real estate lies in its wildlife. Even land that is quite productive on a commercial basis for food or fiber has a growing value as a home for wildlife. Buyers today are often more interested in the quality of habitat for the Cactus Wren than the carrying capacity for Angus cattle. Consumptive uses such as hunting are still very important in many areas, but are now joined by nonconsumptive uses such as bird watching. Commercial exploitation of these values is growing as well, as landowners learn that there is low-overhead income to be derived by merely permitting access.

As a rural real estate professional, it is your responsibility to learn about the wildlife in your area and particularly on your listings. Visit with wildlife professionals to learn of the various animals present in the area, their abundance, and their natural history. Stay abreast of commercial ventures involving wildlife including, but not limited to, hunting, fishing, observation, and interacting. Because of the transient nature of many wildlife species, it is important to know the history of the wildlife populations on your listings as well as on the neighboring properties. In smaller tracts, landowners "share" wildlife, so activities of neighboring owners directly affect the wildlife, the enjoyment, and therefore the value of a piece of ground.

Commercial hunting truck. Note dog cages and raised seat in back.

Unless your listing is plowed ground from fence to fence, it has a wildlife value. Even a monoculture of grasses will support some wildlife for some time during the year. While perhaps the most incorrectly used phrase in selling rural real estate is "great hunting," most properties do have hunting value for some species, whether the consumer is "hunting" with a gun or a camera. Always keep in mind that, for the most part, wildlife presence on a tract of land evolved there and requires little assistance to remain. The risk of reducing wildlife populations is highest when the habitat is adversely altered or wildlife populations are improperly harvested.

Like cattle, wildlife have dietary requirements that cover all four seasons. If an animal is present and reproducing, these requirements are being met at the minimum. Removing or reducing one or more components of the animals' environment may cause the animals to move out when that component is needed and lacking. Additionally, animals require a certain amount of physical cover or concealment to feel safe, otherwise the animals will not remain.

For these reasons, it is necessary for the practitioner to become conversant in the species present and the animals' natural history. If turkey inhabit an area in which you have a listing, you should know the food habits of turkeys, vegetative preferences, daily routine, annual life cycles, breeding season, and hunting season. The same would apply for all game species with additional information on nongame species such as migration periods for birds.

Commercial hunting or outdoor experience operations are fast becoming popular and profitable. The dude ranches of the fifties have been replaced with slick operations ranging in services from merely space to camp out to gourmet meals in magnificent lodges. The practitioner's familiarity with these types of income producing uses for rural land is important whether selling raw land to be developed or reselling an existing operation. If you are aware of any such resorts in your area, you should visit them and learn what is offered, where clientele is derived, and the cost for services. Oftentimes, profitable hunting operations can be developed on small acreage, using pen-raised game; combined with an outdoor experience, this can be a source of income during the off-season.

Farming

The most complex use of rural land in the United States is farming. Gone are the simple farmers of old, plowing two-acre fields with a mule and a one-blade turn plow. Today's farmer is most likely college educated with an equipment inventory valued at six figures. Certainly the small farm exists, but the techniques of farming are big-time. Additionally, gone are the days of one-crop fields with brush edges and fallow periods. Today's farmer sees empty land as nonproductive land. These farmers plant right to the edge of the field (unless they are smart enough to include wildlife habitat in their program) and harvest three crops a year. In addition to corn, wheat, and sorghum, today's farmers produce everything from mushrooms to Christmas trees, apples to kiwi fruit, pecans to almonds, and cotton to guar.

As a real estate professional, your job is not to tell someone how to farm, just as you should not tell a rancher how to run cattle. The practitioner's job is to learn the requirements of the buyer and fulfill such requirements. The ecological requirements of farming are such that rarely will a buyer come into an area to start a farming operation that is not already in the area. Soils, weather, markets, and support businesses are too important to farming for much "wildcatting" to occur.

Therefore, practitioners should become proficient in discussing water, soils, weather, and agricultural production in their area. Additionally, to work with prospective farmers, the practitioner should locate well drillers, earth movers, farm implement dealers, seed dealers, fertilizer sellers, and produce markets, and know the costs of fuel and electricity. In rural real estate, having a listing "near good schools and shopping" just will not cut it. The practitioner needs to make farmers feel comfortable that not only will they be able to practice their craft, but will have the support network necessary to do so competitively.

Environmental Considerations

As the real estate agent representing the seller or the buyer in rural real estate, the practitioner has new additional responsibilities regarding toxic wastes and contamination. Because of various federal acts, including the Comprehensive Environmental Response, Compensation and Liability Act (CERCLA), the practitioner may be liable for losses in value or site cleanup costs related to hazardous waste found on a property in which the practitioner participated in the sale transaction.

Problem areas in rural real estate are old farm or ranch dumps that contain residue of herbicides or pesticides that once were legal and commonly used, but now are considered hazardous waste; barns or sheds where such materials were stored or mixed; underground or above ground fuel storage tanks; vehicle storage sheds or yards; discarded equipment with residues in applicator tanks or booms; and structures with asbestos.

For the practitioner's own protection, if there was ever, to your knowledge, any human use of the property, it pays to get a Phase I Environmental Site Assessment. These are prepared by professional environmental inspectors and consist of a review of past practices on the tract, any files held by regulatory agencies, and a visual inspection of the tract and adjoining properties. The key phrase here is "to the practitioner's knowledge." As a professional, part of the practitioner's listing process should include a detailed interview of the present landowner. Part of that interview should address chemicals of any type ever used on the property and should include questions on when and where used, stored, or mixed, and where the empty containers were disposed. Any indications that any chemicals, batteries, asbestos, fuels, or other wastes were disposed in any fashion other than presently approved, should prompt you to ask for a Phase I assessment.

The practitioner's visual inspection of the property is also important. As the property is inspected, look for newly-turned soil, dead vegetation, old roads seemingly going nowhere, and old equipment. Inspect barns and sheds and make note of old cans, barrels, spray equipment, odors, or stains on the floors. One ranch that recently sold had a new dump that the owner buried to "clean the place up." Upon questioning, the landowner said that the dump was where he regularly burned car batteries. When cooled, he sifted through the ashes to collect the lead as a source of sup-

Bare ground or newly-turned soil may indicate a buried dump

plemental income! Naturally, the dump was dug up, the soil was removed to a depth of fifteen feet, and the soils hauled to an EPA-approved dump site for lead.

Do not trust the seller to know what is or is not toxic and be sure to ask about old dumps. One landowner had owned a property for twelve years, but there was a dump used by the previous owner that contained DDT and PCBs. The present owner chose to dig a new dump upon purchase of the property, and the road to the old dump was completely overgrown. Without careful questioning, it would have remained and cleanup costs could well have been shared by the broker.

What about those instances where nothing is visually evident and the landowner or seller chooses not to disclose a waste site? If you have done your best to learn of any possible contamination of the property, advised the landowners of their potential liability for failure to disclose such contamination, and had any questionable areas inspected by an environmental inspector, your liability is limited. However, there is always the possibility that the other side will claim that, as a real estate professional, you "should have known." It is a very dangerous area of real estate. Practitioners must be aware of the regulations and their responsibilities.

In some areas of the United States, there is another federal regulation that is important to rural real estate professionals. This regulation is the Endangered Species Act (ESA). The ESA is presently under revision by Congress, but the ESA presently states that if a tract of land, of any size, contains habitat or individuals of a species currently listed as endangered, neither that habitat nor the individuals can be harmed in any way. This has been interpreted as meaning that there can be no clearing of the land, no tilling of the soil, no impoundment or draining of water, nothing that materially changes the land so as to make it unsuitable for the endangered species, even if the land is used by that species only for a short time each year.

Areas and species affected by the ESA are usually well known to locals, landowners, and other real estate professionals. However, when taking a new listing, particularly in an unfamiliar area, you should call the United States Fish and Wildlife Service to ask if there are any sensitive areas in the region. If so, ask if any of the listed species occur or have occurred in the vicinity of the listing, and ask for an inspection by the United States Fish and Wildlife Service. An inspection of the property will be made by the United States Fish and Wildlife Service and a letter will be issued stating that there are no endangered species that will be affected by land use practices on the subject tract or listing restrictions of the land's use under the ESA. Another agency that is helpful in this regard is the Natural Resource Conservation Service (NRCS), as well as state wildlife agencies.

SOURCES OF INFORMATION

Not having a rural background does not preclude your being a successful rural real estate professional or selling rural properties. There are many sources for the information you will need to effectively market rural real estate. Several sources are discussed in this section.

Governmental Agencies

Easily the best sources of information on rural land are governmental agencies. These agencies exist solely to assist agents and buyers in assessing the land in terms of its productivity, both consumptive as well as nonconsumptive. At the federal level, the Department of Agriculture and the Department of the Interior have field offices in every state that are storehouses of information. These offices will provide information to agents and to the new owners free of charge in the form of literature, advice, and written management or development proposals. The federal agencies are ideal sources of information on weather, environmental regulations, marketing of agricultural produce, soils, and conservation techniques. Additionally, there are numerous programs administered by federal agencies that will assist landowners financially in the development or management of their new acquisition.

State agencies are perhaps more helpful with regard to specific questions and specific practices in the area. In particular, many states have extensions of their agricultural colleges in each county or parish. The agents of this service, called county agents, are in place to disseminate current scientific research in the areas of farming, ranching, and rural land management. States also maintain agencies concerned with water, rural product marketing, wildlife, and the environment.

One of the best things about government agencies is the free literature these agencies produce. This literature covers everything from fruit tree management to treating cows for flies. It is clearly written and is precise enough to be used by the novice. Agents should learn where these agencies' offices are located and visit their offices frequently, if for no other reason than to stock up on free pamphlets that can be passed on to prospects and buyers.

Personal Contacts

Another excellent source of information on any rural property is the seller. It pays big dividends to spend time with the seller, learning about past practices, weather, and the history of the land. Local agricultural professionals, suppliers, and "old-timers" will often happily share their knowledge about the area and may have information about your listing in particular.

Finally, your personal reference library should include the county soils books for the counties you operate in. These are available from the Department of the Interior. From the United States Geological Survey (USGS), you can obtain topographical maps that are invaluable for showing aspects of relief on your listings. Agents need a good county map, showing roads, waterways, and towns in a scale that is easy to read. Some of the helpful text resources available include *Ecology and Field Biology* by Robert Leo Smith (Harper and Row, New York); *Farm Management* by Ronald D. Kay (McGraw-Hill Company, New York); and *Range Management* by Lawrence Alexander Stoddart, Arthur D. Smith, and Thadis W. Box (McGraw-Hill Company, New York). Most of all, you need the good graces of other rural real estate professionals who will assist with comparables, changes in laws and regulations, trends, and marketing techniques peculiar to rural real estate. Because of the myriad of variables in rural land, cooperation with other agents is always a good policy.

12 Development Plans

CASE HISTORY

The following is an actual development plan for a ranch in South Texas. While it is very specific to that area, a great deal can be learned from it by the real estate professional. Make note of the treatment given to the ecological aspects, weather, soils, water, and flora as well as the physical components. Much of the information gathered here should be part of the listing process.

Knowing the potential of the property, be it in wildlife, farming, or ranching, will help the practitioner to effectively market a property. As already noted, nothing turns a buyer off faster than an agent who claims every property has "good hunting" or is "great farmland." Learn what the buyer needs to know and be aware of the surroundings.

General Information

The Smith Ranch is located 9.5 miles southeast of Charlotte. Access is gained using State Road 97 from Jourdanton to the east or on 173 from Devine to the northwest. Major highways in the area are U.S. Highway 281 through Pleasanton, 24 miles east northeast of the ranch on State Road 97 and I.H. 35, 32 miles northwest on farm to market roads at Devine. Towns in the area of the ranch are, of course, Charlotte, Jourdanton (the county seat), Pleasanton, Devine, and Christine.

Of these towns, Pleasanton, founded in 1858 as a major cattle trail origin, is the largest with all the services of a larger town. To the west of Pleasanton on State Road 97 is the regional airport, located 22 miles from the ranch. Pleasanton has a number of motels and restaurants and is the business and social center of Atascosa County.

Jourdanton, as mentioned above, is the county seat. Although Jourdanton is substantially smaller in population (2,299 versus 6,000) than Pleasanton, it still provides the necessary services to area ranches. The county hospital, Mercy Hospital, is in Jourdanton as are the sheriff, emergency medical service, the county agent, and the livestock exchange. The power company and the telephone company have branch offices in Jourdanton.

Charlotte, once a boom town, has steadily declined over the years and no longer provides anything more than the most basic supplies/services to area ranches. There is no hospital in Charlotte, nor are there any implement companies, feed stores, or craftsmen. It does have two cafes, one grocery, a good lumberyard/hardware store, one automobile dealer, a branch of the Atascosa County State Bank, and a convenience store. The post office is located one block east of Highway 140 in "downtown" and is a full-service facility. Area services are provided in Fig. 12-1.

Atascosa County

Agriculture. According to the 1990 Texas Conservation Needs Inventory, approximately 488,386 acres in Atascosa County were in rangeland, 217,379 acres were in cropland, and 38,072 acres were

Figure 12-1

AREA SERVICES

County Agent: Mike Schook
323 Circle
Jourdanton, Texas 78026

Soil & Water Conservation Service:
(Natural Resource Conservation Service) Pat Mudd
123 Water Street
Pleasanton, Texas 78334

Agriculture Stabilization & Conservation Service:
(ASCS) Moe Money
4356 Copper Lane
Charlotte, Texas 78801

Implement Companies: Joe Shady
19 Broken Wallet
Poteet, Texas 78332

Frank Crooked
456 High Dollar
Jet, Texas 78210

Veterinarians: Dr. Will Fixum
999 Post Oak
Tilden, Texas 78208

Dr. Les Vacas
111 Elm
Charlotte, Texas 78231

Livestock Supplies: Tightwad & Company
436 Main
Pleasanton, Texas 78324

Feed Stores: Joe Penneyless
438 Elm
Pleasanton, Texas 78238

Billy Bankrupts
567 Tornado
Pleasanton, Texas 78238

Lumberyard: Not in Stock Lumber
438 Broadway
Charlotte, Texas 78324

Gas & Oil: Big American Oil Company
9087 Foster, Suite 46
Jourdanton, Texas 78320

Dozer Service: Virgil Treeknocker
351 Barren Plains
Charlotte, Texas 78209

improved pasture. Approximately 29,066 acres of the cropland were irrigated. The main crops grown are grain sorghum, wheat, oats, peanuts, corn, watermelon, and flax; in some areas vegetable and fruit crops are grown as well. Agar, a plant used in a variety of industrial processes, grows well in Atascosa County, but recent cutbacks in contracts stopped a "bloom" of that crop. The livestock operations are mainly cow/calf; there are some stocker cattle operations. Supplemental feeding is needed for short periods in winter and during droughts, which limits the stocker operations.

Natural Resources. The main source of irrigation water is the Carrizo Sand Aquifer, which varies in depth throughout Atascosa County and borders the Edwards Aquifer on the north. Oil and gas are major resources, mainly in the western part of the county. Uranium has been mined in the county for several years. Lignite is present in the county in commercially valuable deposits and sand is mined in the northern part of Atascosa County for sale in San Antonio.

Climate. In Atascosa County, summers are hot and winters are mild. Cold spells and snowfalls are rare. Rains usually are heaviest in fall and are associated with dissipating tropical storms.

Precipitation is usually adequate for range vegetation. However, because of the high rate of evapotranspiration, it is commonly not adequate for cotton, small grains, or vegetables; supplemental irrigation is generally needed for these crops. The total annual precipitation is approximately 28 inches. Of this amount, 17 inches or approximately 60% usually falls during April through September. In two years out of ten, the rainfall during April through September will be less than 5 inches.

Ranch History. Based on visits with the local population, it was learned that the present 1,600 acres under survey was once the headquarters for a much larger Home Ranch. Mr. Sam Smith and his wife, Mary, combined their properties to form one ranch which was in excess of 10,000 acres. Mr. Smith died in 1967 and his livestock and the implements were auctioned off at that time. From 1967 to the present, the ranch has been sold off in pieces until all that remains is this 1,600 acres and the 300-plus acres across the road from the front gate. These 1,900 acres were leased by Mr. David Blue and Associates from the Smith family until purchased in 1990 by the present owner. Mr. Blue made the improvements and changes in the headquarters that exist today. The present owner did not make any improvements on the ranch and the ranch has not been grazed since Mr. Blue removed his cattle in 1991.

Neighbors. Based on information received from the county tax office as well as the Agricultural Stabilization and Conservation Service (ASCS), a list of the surrounding neighbors has been compiled. This information, with the respective addresses, is provided in Fig. 12-2.

All neighbors are absentee owners except Mr. Blue, the only live-in neighbor. Mr. Blue is located one and one and a half miles toward Charlotte on the left, and he also has the nearest telephone. Mr. Blue is agreeable and has offered his assistance in exchange for some wildlife management advice.

Figure 12-2

NEIGHBORS

Starting on the road at the ranch's northwest corner, and moving east, the following is a list of the neighbors. The code number to the left of each name is the property designation as recorded at the ASCS office in Charlotte.

D87 I. M. Petty
543 Roads Lane
Moore, Texas 76584

D98 Buster Phideaux
298 Linda Lane
San Antonio, Texas 78209

D96 Fred Guthery
39087 Still
Boston,
Massachusetts 02116

D93 Bob Moxie
768 Hill Street, Apt. #3
Franklin, Tennessee 38138

D92 David Blue
Southland Ranches, Incorporated
435 South Park
Austin, Texas 78701

Present Situation

Fields. See Fig. 12-3 for field and pasture designations and Fig. 12-4 for breakdown of soil types.

Soils. The ranch is comprised of twelve soils ranging from approximately twenty-eight percent to less than three percent of the total. The "front" or cultivated portion of the ranch shows less variety in soil types and consists primarily of Coy Sandy Clay Loam in the fields, Monteola Clay in drainages, and Campbellton Loam on the high ground. Field #1 as depicted in Fig. 12-3 is the most uniform, consisting almost entirely of Coy Sandy Clay Loam while field #6 is almost as uniform combining Monteola Clay with the Coy Sandy Clay Loam. Figure 12-4 shows a breakdown of the soils and gives evidence of prior plan-

Figure 12-3. Field and Pasture Designations

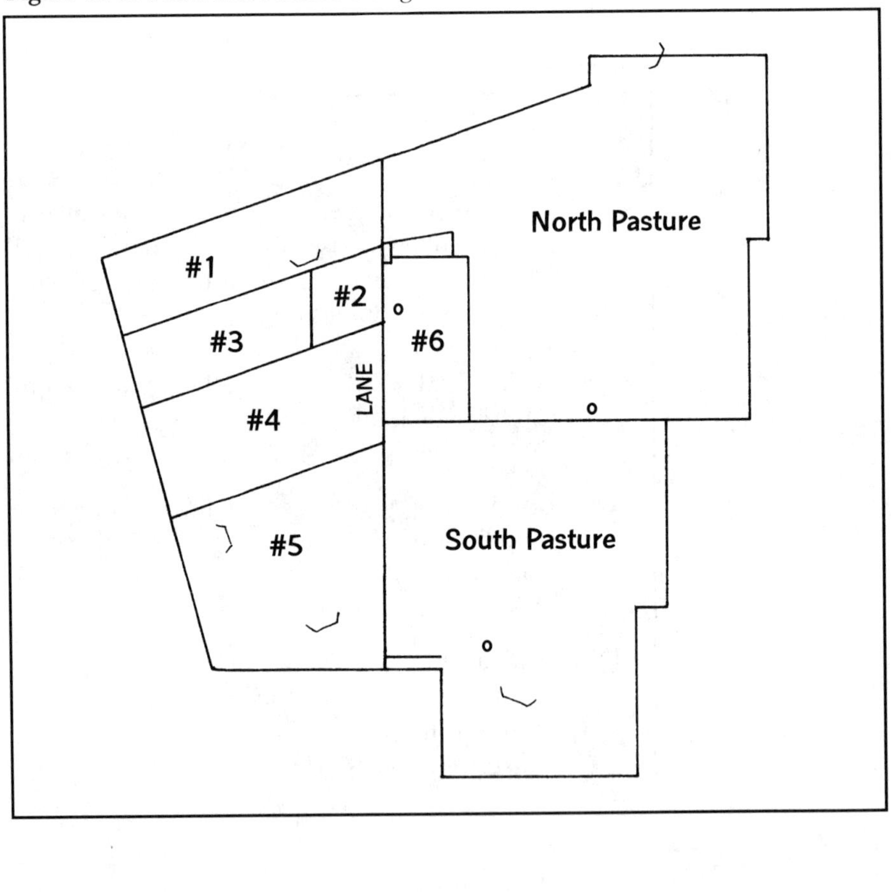

Figure 12-4. Soil Types

S&W Soil #	*Water Avail*	*Run-Off*	*Perm*	*Eros Haz*	*IRR Crop*	*Non-IRR Crop*	*IRR Past*	*Native Past*	*Wild Habit*	*% of Prop*	*Notes*
2—Amphion Sandy Clay Loam 0–1% slope	High	Slow	Mod Slow	Slight	High	High	High	High	Med	5%	Highly Corros
3—Amphion Sandy Clay Loam 1–3% slope	High	Med	Mod Slow	Mod	High	High	High	High	Med	3%	Highly Corros
4—Campbellton Loam	Med	Med	Mod Slow	Mod	Med	Med	High	Med	Med	10%	$CaCo_2$
6—Christine	Med	Slow	Slow	Slight	Med	Med	Med	Med	Med	10%	Mildly Corros
7—Coy Sandy Clay Loam	High	Med	Very Slow	Mod	High	High	High	High	Med	28%	Slow Perm
11—Elmendorf-Denhawken	High	Med	Very Slow	Mod	High	High	Med	Med	High	8%	Slightly Corros
20—Imogene Fine Sandy Loam	Low	Slow	Very Slow	Slight	Low	Low	Med	Med	Low	10%	Very Salty
22—Laparita Loam 0–1% slope	Low	Slow	Mod Slow	Slight	Med	Med	High	Med	Med	3%	Mod Salty
23—Laparita Loam 1–3% slope	Low	Med	Mod Slow	Mod	Med	Med	Med	Med	Med	5%	Very Salty
25—Monteola Clay 0–1% slope	Med	Slow	Very Slow	Mod	High	High	High	Med	Med	5%	Corros
26—Monteola Clay 1–3% slope	Med	Med	Very Slow	Mod	High	High	High	Med	Med	10%	Corros
33—Papalote Fine Sandy Loam	Med	Slow	Slow	Slight	Med	Med	High	Med	High	3%	Corros

shows a breakdown of the soils and gives evidence of prior planning in the location of the fields and pastures of the ranch.

Plants. Field #1, 133 acres, is perhaps the cleanest field, and has a cover of native grass. There are mesquite trees scattered across this field and a small mud trap on the entry road. Field #2, 22 acres, has a combination of dry land willow, coastal, and was at one time planted with plains lovegrass. The field slopes down from the main well to Field #3 where mesquite brush has begun to spread. Field #3, 60 acres, has been planted in coastal and has fairly good cover along the two drains which feed the big tank to the south. Field #4, 109 acres, was planted for about two-thirds of its width in coastal and the remaining one-third was recently planted in sorghum which never matured. The coastal is spotty and needs cultivation as well as fertilization. Field #5, 206 acres, contains the big tank. It shows no signs of recent cultivation and the brush is coming back throughout. Field #6, 99 acres, is perhaps the next cleanest and is presently covered with a mixture of native grasses and sunflowers. It was planted in oats to attract deer and has cover on three sides inside the fence.

Field Fences. The fences of all the fields are in good condition with the exception of the ranch perimeter fence from the entrance to the southwest corner. This fence is overgrown and has been patched a number of times. The wire itself is marginal and would not stand up to being restretched. The posts are in poor condition as well, yet the fence is "whole" in that it would contain livestock. Throughout the fields there are "gaps," gates made of wire that weaken the fences and should be eliminated wherever possible. Field #6 is bordered to the west by the old lane consisting of a combination of barbed wire, chain-link, and netting. This should be replaced and will be if the lane is reestablished (see proposed improvements).

Topography. The fields, as previously stated, show a good deal of planning as per soil types. The nature of these soil types also is indicative of the topography of the fields in that the fields all slope down from the ridge in the center of the ranch to the drainage flats below. Fields #1 through #5 slope from the perimeter to that same drainage flat, giving the system of Fields #1 through #5 a gutter effect feeding into the big tank in Field #5. Slopes are gradual and provide good permeability to the soil which is a plus since, as illustrated in Fig. 12-4, both the Coy and Monteola soils have slow permeability and a moderate hazard of erosion.

Pastures. (Fig. 12-3)

Soils. The soils of the rangeland of this ranch consist of all twelve soils mentioned previously and listed on Fig. 12-4. The course of the Macho Creek through this area has formed layered deposits of the various soil types as well as "islands" of distinct soils. Referring to Fig. 12-4, these "alluvial" soils are of poor quality and are not suitable for cultivation although most are suitable for native range or nonirrigated pastures.

Topography. The topography of the rangeland consists of two rolling ridges which run north to south parallel to the Macho Creek. The first ridge forms the midline of the ranch and has the highest elevation in the immediate area. Down from that ridge east three hundred yards, the second ridge lies along the Macho Creek. The Macho Creek itself is a confusing network of gullies covering one-sixth of the ranch's total area.

Fences. (Fig. 12-5) Here again, the interior fences are in good shape while the perimeter fences are marginal. From midline in the north (headquarters) and moving east, the fence is well mended and both the wire and posts are aged but sturdy. The eastern perimeter fence, from the northeast corner to the southeast corner, is badly overgrown and the wire in places is loose, stretched out, and old. The posts are old yet still usable and the entire length needs tightening and repair before stocking. The fence from the southeast corner to Field #5 is marginally better but severely overgrown in the vicinity of the Macho Creek. The creek itself presents a problem of the necessary water gaps in the gullies. The existing gaps are old and in poor shape both physically and in design.

Figure 12-5. Condition of Fences

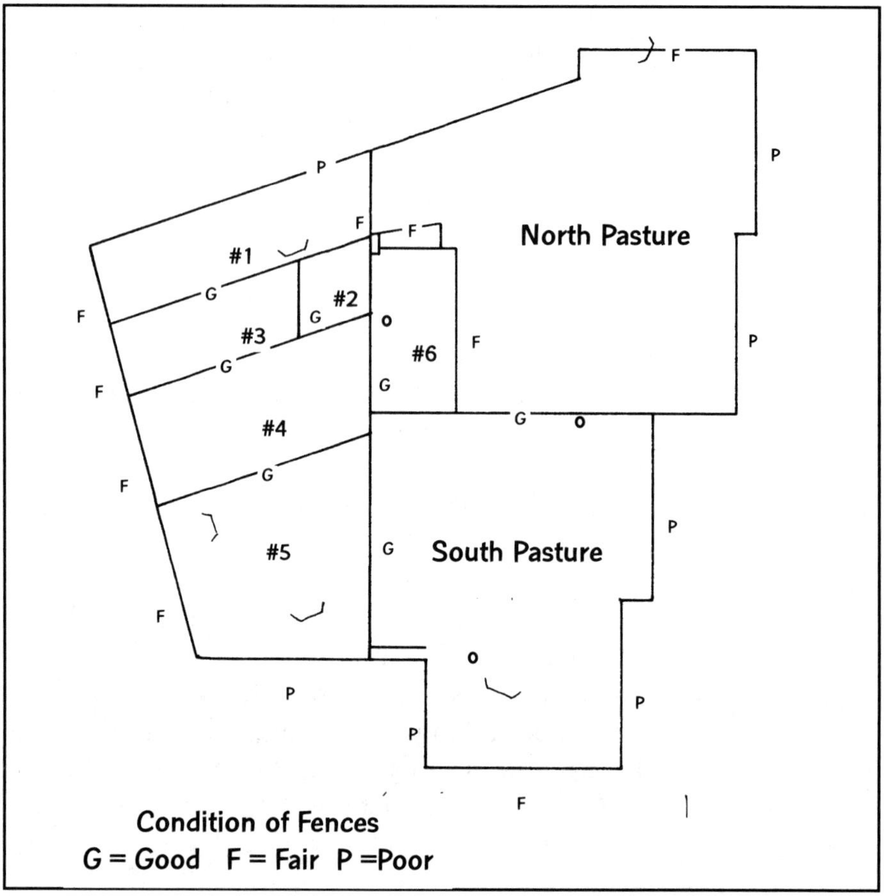

Plants. The flora of the pasture naturally is determined by the soil type and the availability of water. Along the midline of the ranch, Campbellton Loam and Coy Sandy Clay Loam have the water availability to support blackbrush and the populations of that plant species follow the soil delineations exactly. Moving onto the second ridge, mesquite and whitebrush dominate due to the higher density of the soil which allows less permeability and water availability. In the smorgasbord of soils along the Macho Creek, live oak, hackberry, mesquite, whitebrush, persimmon, and willow are found because the drain provides abundant water at times while the heavy clay soils retain water longer allowing for the establishment of trees.

The only other soil association that makes up more than 10% of the ranch is a strip of Elmendorf-Denhawken soil combined with Imogene Fine Sandy Loam. This strip is located roughly in the low ground between the two ridges and supports whitebrush thickets.

Native grasses throughout the pasture are in good shape due to the layoff of the past sixteen months. A great number of species exist with sideoats grama, Texas cupgrass, and pinhole bluestem making up 30% of the total. Forage provided by these native grasses is good and the variety determined by the different soils and slopes ensures continuous availability with planned grazing.

Water. (Fig. 12-6) Water on the ranch is available from the two sources common to South Texas, wells and stock tanks. Although the Macho Creek at times carries a great deal of water, it is primarily a flood drain and cannot be considered a source of water.

Stocktanks. There are five stocktanks on the ranch, one of which is shared with the neighbor in the northeastern corner. All five tanks are shallow and have silted in a great deal. Tank #1 was built to provide water to the area around Field #1 as well as to catch silt moving toward Tank #3 (big tank). This was also the purpose for tank #2, which lies above tank #3 near the western perimeter. Both tanks #1 and #2 are dry as of this writing. Tank #3 covers potentially twenty acres and is the largest tank in the immediate area. It was built on a natural flat with little clearing of the projected tank bed. Stumps and snags abound. The tank has silted in from a potential depth of eight feet to its present potential of about four feet. As of this writing, it is fourteen inches deep and covers approximately eight acres. Tank #4 dams the Macho Creek and appears to be the newest tank. Although silted in, it can still hold a great deal of water with a potential of about eight acres at a depth of six feet. This tank is almost dry. Finally, Tank #5, which is on the northeast perimeter, is a one-half to one acre tank which has silted in to the point that it is no longer a dependable water source. This tank is also dry at this time.

Wells. (Fig. 12-6) The ranch has three known wells, two of which are not currently in use. The windmill (Point C on Fig. 12-6) was blown down and the blades and motor were damaged. It is a shal-

Figure 12-6. Tanks and Wells

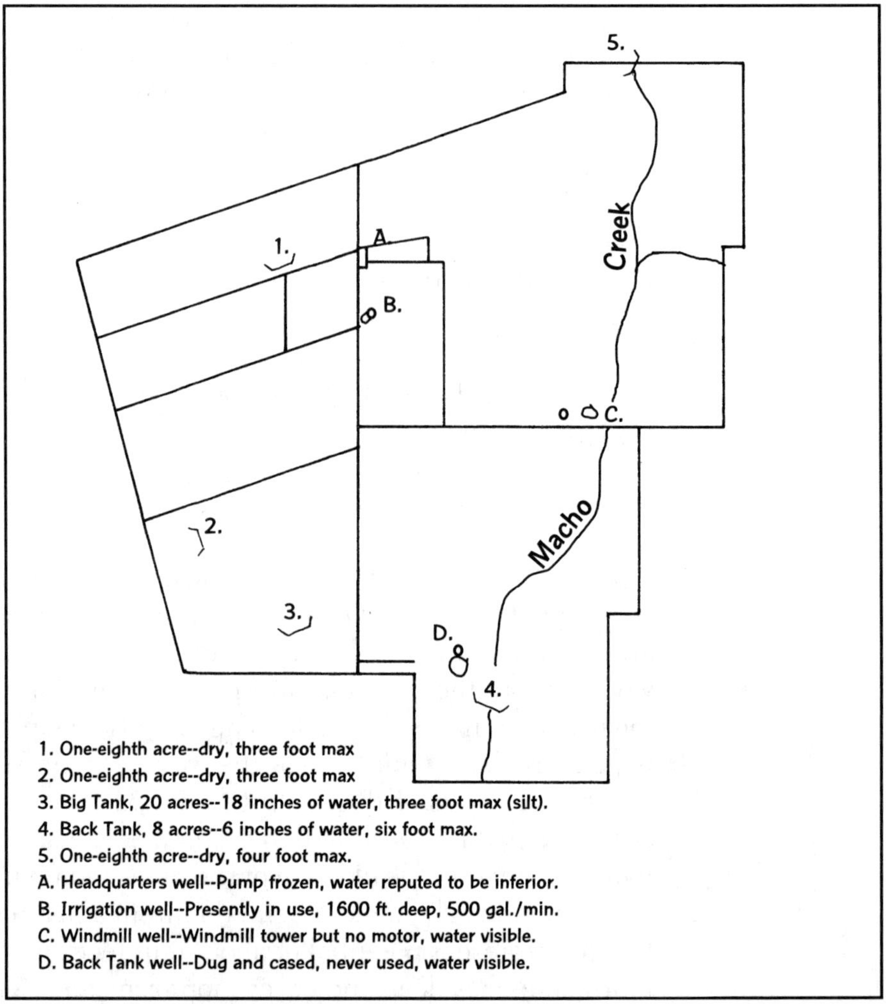

low well according to Mr. Blue. Another mill is found at the dump which seems to indicate that the well is dependable enough to repair and has been repaired in the past. The headquarters well, (Point A on Fig. 12-6) has a Dempstar belt-driven pump on it which is frozen from disuse. Although the pump's capabilities are unfamiliar and unknown, Mr. Blue said that the well was shallow and the water was of poor quality. This well is not in use and has been bypassed in the headquarter's water system. It has a work-over tower which is now unusable since the tower has become the focus of the headquarters' power lines. The irrigation well (Point B on Fig. 12-6) had a seven and one-half horsepower electric motor driving a Peerless wellhead, but is now used with a submersible pump. The well has been used in the past for irrigation with the above-mentioned equipment and produced four hundred plus gpm. A flat storage tank (earthen) is located at this well as is a selection of four inch pipe used in irrigation. A canal extends north from the well to the headquarters and indicates flood and siphon-pipe irrigation. The submersible now in use is attached to a 200-gallon pressure tank and supplies all the available water. Mr. Blue said that this well was eighteen hundred feet deep into the Queen City strata and provides dependable water. It was found that when the water is allowed to stand, the sulphur salts will settle out and the water becomes more palatable. A Department of Health water sample was submitted and the results will be attached to this report. This well, too, has a functional work-over tower.

Physical Plant. (Fig. 12-7)

Houses. The ranch has three houses by design and use. All three houses have carpeting throughout, gas outlets of various types, bathrooms, and porches. At the time of this survey, all three were watertight and the tin roofs were intact. The houses were built on raised (wood) foundations with asbestos siding. Most of the windows are aluminum with coverall screens (no frames). All were remodeled to the present configurations in the 1979–1990 period by Mr. David Blue and Associates.

The big house (Building A on Fig. 12-7), consists of a bunk room, living area/kitchen, bathroom, and a utility room with hookups. The house has five ceiling fans, cabinets and sink, wrought iron on the windows, and both a front and back porch. The house has a good deal of space, both living and storage, and was intended to be used as the common area for hunters to eat, play cards, etc. The water pipes are in poor condition and there is a leak under the bathroom. This house has a large, electric water heater which appears to work well. The house also has the closest thing to a yard, with planted trees (oak), shrubs (oleander), and carpet grass. There are also yard lights intended for evening entertaining.

Figure 12-7. The Headquarters Complex

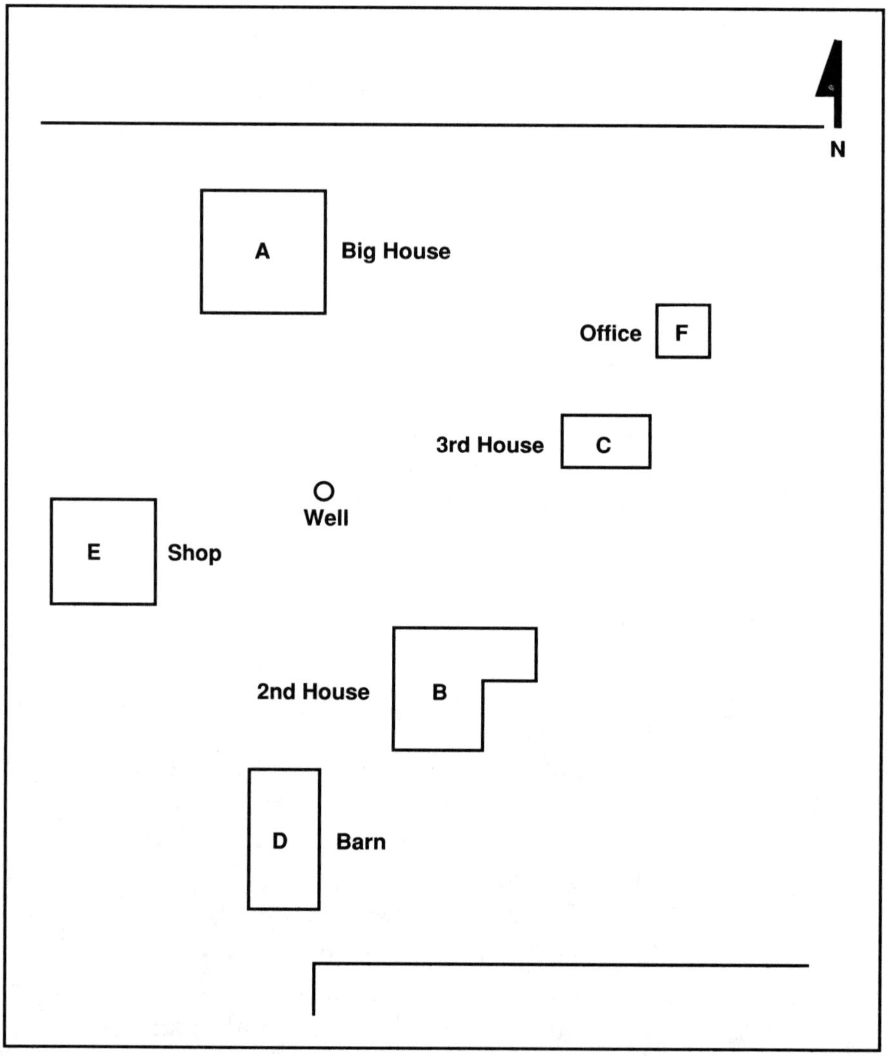

The second house (Building B on Fig. 12-7), consists of four equal sized rooms, a bathroom, and a hallway. This house has both front and back porches although the back porch is in need of repair. This house needs work on the windows as well as screens on the windows, and rehab of the bathroom. A major leak exists under what was the kitchen and water to this house has been shut off for that reason. Some panels of siding are missing and the front porch sags a bit. The house has two large closets and some curtain hardware.

The third house (Building C on Fig. 12-7), consists of two equal sized rooms, a bathroom, and a hallway/dressing room. This house is air-conditioned and has small porches front and back. This house has no water that could be found. Apparently, the water originally came from Building B and the corrosiveness of the soil eroded the pipe. There are some broken windows and screens are needed. The house sags in the middle, but this has not affected the doors or windows. The bathroom of this house also needs work.

Barns and Sheds. The ranch has three other buildings which have been labeled the barn (Building D), the shop (Building E), and the office (Building F).

The barn is a twenty by forty foot wooden structure with a tin roof and a shed built along the western forty-foot side. The barn has sliding doors at both ends and a dirt floor. The barn is watertight and is estimated to hold approximately 2,000 bales of hay. The structure is sturdy and needs little repair except for rehanging the doors.

The shop is a twelve by forty foot building with sheds on both long sides. It is built on an elevated (wooden) foundation and has one room, three doors, and a number of windows. The shop is wired for 220, has water outlets for a washing machine and dryer, and has its own water heater. The shelving suggests it was used as a shop as do the sheds along the sides. The shop needs quite a bit of work, but is structurally sound.

The office is a twelve by six foot building constructed on an elevated (wood) foundation with shed all around. The office is paneled, and has extensive shelf/cupboard space. It has 220 wiring, a vent for a wood stove, and a slot for a wall air conditioner. The office is watertight and does not need repair.

Pens. There are, as such, no livestock handling pens on the ranch although a small trap lies along the northern edge of Field #6. Also, there are small traps running down the midline fence from the headquarters to Field #5.

There are four gravel test sites on the ranch, three of which have been used at one time or another as dumps. There is a dilapidated shed on the southwest corner of the big tank which can serve no function. The numerous ground deer blinds are in good condition, but the one tower blind located in Field #5 is in poor condition. There is a game cleaning table between Buildings B and C.

Wildlife. Perhaps the most important game species, in terms of hunting dollars spent, in Atascosa County is the bobwhite quail. The Smith Ranch lies in prime country for this bird and is suitable for the Blue (Scaled) quail as well. Whitetail deer, although larger than those in central Texas, are not of the trophy class unless intensively managed. Javelina, feral hogs, cats, dove, and rabbits add to the potential profit from lease hunting in Atascosa County.

Quail. To attempt to determine the number of quail on the ranch, the strip census method developed by O. E. Frye was used with trained pointers. The process was further refined by choosing transects on both the dry (northeast) and wet (southwest) portions of the ranch so as to determine the success of this year's hatch. The transects were run between six and eight o'clock in the morning while it was cool and the wind, common during the three weeks spent on the ranch, had not yet started. Transects were measured out on a straight line along an easily defined feature (fence, road, etc.) and the distance from this line for each point was computed giving the area surveyed as well as the number of birds in that area. The dry transects had a number of solid points, but on flushing, it was found that the birds were still paired up for the most part. This late pairing and the absence of young birds indicates that the water sources in this area of the ranch are insufficient to support the nesting period. Points and flushes in the wet transects showed not only full coveys, but both young and newly hatched birds, evidence of a double hatch. Results of the census support a total number of coveys at thirty, each containing an average of twelve birds. Blue quail are not easily counted; however, three coveys with an average of five birds each were seen.

Deer. Considering the size of the ranch and the lack of game fencing, the survey of the deer population was concentrated on the condition of the herd as opposed to total numbers. Using both the spotlight and Hahn Cruise methods, it was discovered that due to the intense water shortage, the deer have congregated around the big tank in Field #5. This is to be expected, but it made census taking inaccurate due to the "yarding up" of the deer in thickets east of the tank. An aerial census, although it would be more accurate, would be affected by this as well, and considering the limitation mentioned above would not be economical. Therefore, most of the time allotted to deer was spent observing the deer from blinds and at the tank using a light. The deer are in good shape with smooth pelts and are alert. The sex ratio is approximately one to five based on these periods of observation and the bucks are obviously the rejects from years of improper hunting. Only one good buck, two inferior bucks with small antlers, and one long spike buck were seen. Yearlings and fawns were observed, but the combination of predation and the drought has reduced the fawn crop to 25% or one fawn per four does.

Other Wildlife Dove were extremely abundant during the survey period. The birds cross the ranch from north to south with the majority gliding on the ridge at mid-ranch. The big tank is a natural water stop and will provide excellent hunting. The sunflowers in Field #6 are attracting quite a few birds and could be hunted as well. Rabbits, both cottontail and jackrabbits, abound but are presently congregated near the tanks and are difficult to hunt. No evidence was found of feral hogs at the big tank, and upon investigation it was learned that the Smith Ranch has been a longtime favorite hunting area by locals who chase the hogs with dogs. There is a population of javelinas in the northeast corner that lay up in one of the gullies near the old dump. Although the big tank is full of carp, fish per se do not exist on the ranch due to the unreliability of the tanks. The big tank also supports one of the highest concentrations of turtles, both soft- and hard-shelled, that has been surveyed. The proximity of the ranch to Mexico gives it extra color in terms of birdlife. Not only was a Caracara (Mexican eagle) observed, but a Grooved-bill Ani as well. A flock of Black-bellied Tree Ducks are summering at the big tank. Coyotes, as earlier mentioned are present on the ranch just as coyotes are found throughout the southwest. Many people mentioned that the area was also famous for rattlesnakes, and although none of the snakes were encountered, it should be remembered that there was a better than average chance of encountering a rattlesnake particularly around the headquarters.

Overall, the ranch has a diverse, healthy wildlife population. With additions of water and the introduction of a brush control program, it should support a great deal of hunting pressure regardless of the changes brought on by a cattle operation.

Proposed Improvements

Fields. It is recommended that a cow/calf operation be established at this time. Cultivation of cash crops does not at present pay for itself with the high cost of farm equipment. At the present time, there is not a well suitable for irrigation of either grain or vegetable crops and the local labor pool is neither skilled nor interested in vegetable harvesting. Fields #2, #3, and #4 have some improved grasses established and the production of those grasses could be easily increased at a low cost. Figure 12-8 lists the minimum tools and implements needed for a cow/calf operation.

The heavier equipment would not need to be purchased, only leased or rented. This would give an initial investment far lower than if an intensive crop farming venture were attempted. This strategy will allow for a faster return to a profit. It should also be remembered that the ranch lies in a low rainfall area with year-to-year uncertainty in precipitation. So, although the ranch will initially be dependent upon the production of one hundred head, with the following improvements and development of the water system, a herd of two hundred could be built. An improved apple or pecan orchard can be planted, commercial catfish can be raised, game can be harvested for profit, and the profitability would not be dependent on the weather.

Cultivation. The fields should be broken deeply after this long a period. Moldboard plows use a one-half plowshare to turn the soil completely over. A sodbuster plow, on the other hand, uses ten to fourteen steel shanks to penetrate the soil and break up the hardpan. This allows moisture to penetrate and air to circulate. Both processes are about the same price ($14 per acre), and can be adjusted up to fourteen inches in depth. Although moldboarding is slower, this method does a much better job of making nutrients available. The effects of a moldboard treatment also last longer

Figure 12-8. Price List

Tools	Cost
Screwdriver Set	3.50
Claw Hammer	11.79
Ball Peen Hammer	9.99
Sledge Hammer	18.95
Post Hole Digger	31.95
Tamping Rod	21.95
Fence Stretcher	30.00
Fencing Tool	12.25
Brace and Bits (Wood)	15.00
Gloves	7.15
Metal Post Driver	12.95
Grinder	85.00
Shovel	21.17
Compressor	175.00
Axe	21.99
Tow Chain	49.90
Lassoes	24.00
Paint Supplies	12.00
Large Pipe Wrench	11.95
Large Crescent Wrench	15.99
Bow Saw	7.99
Hose (50 Ft.)	11.78
Hoe	5.89
Hacksaw	8.49
Tin Shears	9.00
Mattox	20.45
Horse Hoof Tools	23.40
Horn Saw	7.37
Nose Clips	6.90
Burdizzo	38.95

Implements	*I. H.*	*Farm Center*
Tractor—80 hp New	22,800.00	21,400.00
Tractor—62 hp New	15,500.00	15,850.00 (68 hp)
Tractor—80 hp Used	N/A	11,250.00
Cat. (Used)	N/A	55,000.00
Shredder	3,190.00	7,000.00
Blade	1,325.00	1,355.00
Auger	750.00	483.00
Disk	6,500.00	4,633.00
Fertilizer	1,690.00	1,700.00
Rome Plow	N/A	3,500.00
Warranty	2 Years	2 Years
Lease	Yes	Yes
Delivery	Free	Free
Service Charge	$25/hour/.50/mile	$9/hour/.35/mile
Shop	Yes	Yes

which makes it more cost efficient. A third process, rome plowing, should also be mentioned because it has the advantage of being very quick. A rome plow is usually used in brush control and consists of a very heavy offset disc pulled by a tracked vehicle. A rome plow usually turns the top six to eight inches of soil thoroughly, but is more expensive ($30 per acre) due to the expense of the prime mover.

Fertilization. The fields should next be fertilized to replace nutrients lost to past crops as well as leaching. The soils should be sampled to determine the shortages and fertilizer purchased to remedy those shortages. Full nutrient samples are $8 per acre, and will match up available water with the soil sample. The spreader can be leased in Pleasanton, bought, or a custom farmer can be used who charges $3.50 per acre plus fertilizer for the job.

Discing. The next step in field preparation is discing to work the fertilizer in and to break up the clods brought up by the busting process. This is a one-person operation and costs $8.00 per acre on the local market. Discing is a year-round job, though, and it is recommended that the implements be purchased so as to be used off the field as well.

Final Step. The final step in field grass operations is planting. It is recommended that a one pound to twenty-four pound ratio of Kleingrass seed to sorghum or a sorghum hybrid be used to allow the grass time to become established. This mixing serves two purposes:

1. It allows grazing of the field without damage to the young grass seedlings.
2. It also gives protection to the grass from wind and water erosion while still keeping the soil loose.

Straight Kleingrass could be planted and defer the field for a year, but results of both techniques showed that the stand of Kleingrass was both fuller and stronger when started with sorghum.

Time Table of Field Preparations. The sod-busting process should be completed in the fall as should the first discing. Planting should immediately follow fertilization and another discing in January. Field #1 should be first since it is the cleanest and can be easily deferred this fall. Field #2 should also be treated this fall to retard the infestation of dry land willow. Field #3 already has some spotty coastal as does two-thirds of Field #4. These fields should be busted in the spring (March or April) and fertilized. Field #4 should be completely seeded with coastal since Kleingrass and coastal cannot be grazed together (livestock will eat the Kleingrass off first). Fields #5 and #6 should be prepared as was Field #1 according to the schedule in Fig. 12-9.

Brush Control in the Fields. Both Fields #5 and #6 have brush bordering each and it is felt that the brush should remain. The brush in Field #5 provides access to the big tank for wildlife both from the ranch and from across the road. The brush also serves as a shelter for cattle grazing in those fields which saves the expense of building shelters. For the same reasons, it is also recommended that a lane of brush be allowed to establish itself in Field #3. This lane does not need to be wider than 100 feet on either side of the drain and will not adversely affect the productivity while greatly enhancing the quail production in the "front" of the ranch. Along these lines, it would be advisable that the fences be left covered whenever possible as long as the integrity of the boundary remains intact.

Pastures. To effectively utilize the rangeland of the ranch, both from the wildlife and beef cattle perspectives, the rangeland needs to be broken up and have improved grasses introduced. This process serves a number of purposes: It increases the "edge" available to wildlife, it greatly increases the carrying capacity for livestock, it takes grazing pressure off the native plants thereby releasing more food for wildlife, and it facilitates control of both livestock and wildlife.

Brush Control Methods. There are numerous brush control methods available today. These methods can be divided into two broad areas: control through defoliants and control through physical removal of the brush.

Figure 12-9

PROPOSED TIMETABLE

SEPTEMBER THROUGH DECEMBER:

- Lease grazing rights for fifty to seventy-five head
- Harvest deer, quail, and dove per recommendation/leases possible
- Remodel one house for manager
- Build kennels
- Repair water systems in headquarters
- Establish feed/water stations
- Moldboard and disc Field #1

JANUARY THROUGH JUNE:

- Build a lane-type water system with reservoir
- Construct working pens
- Disc Fields #2 and #6
- Fertilize Fields #2, #3, and #4; spring remainder of Field #4
- Stock with one hundred permanent head
- Repair fences as needed
- Plant Field #1

JULY THROUGH DECEMBER:

- Begin brush control program
- Add cross-fencing to establish the eight pastures
- Moldboard Field #5
- Plant orchard where feasible
- Replace front fence
- Harvest deer, quail, and dove as per recommendation

JANUARY THROUGH JUNE:

- Plant Field #5
- Build new stock tanks where recommended/stock with catfish
- Plant brush control strips

JULY THROUGH DECEMBER:

- Moldboard Field #6 (unless in orchard)
- Replace eastern fence
- Harvest deer, quail, and dove as per recommendation

JANUARY THROUGH JUNE:

- Plant Field #6 (unless in orchard)
- Bring herd up to two hundred head
- Fertilize Field #1
- Harvest catfish for commercial sale

The first area concerns the use of defoliants which prevent the plant from maintaining its leaves thereby disrupting the photosynthetic process. Two products recently developed, BROWNOUT and WILT, as well as the old standby, KILLUM, could be used. All three are effective when properly applied and each has its pluses and minuses depending on the plant treated, soils, and weather. WILT is a pelletized or liquid product applied by hand to the base of the plant. This product is expensive considering the labor-hours involved, but WILT is very effective on mesquite. BROWNOUT is granulated and is applied by air. This product is most effective on shallow, rocky soils and does not effectively control mesquite. Use of an airplane is prohibitively expensive on small tracts. KILLUM is sprayed either on the foliage or the trunk (after notching) and is usually mixed with diesel or kerosene. It requires a lot of labor-hours, but is inexpensive to purchase. Of the three products, KILLUM in pelletized form is the easiest and, all things considered, is the cheapest to use. It should be remembered, however, that these methods leave the brush and the area cannot be cultivated with wheeled vehicles.

Physical control of brush consists of five processes: Root plowing, rome plowing, roller cutters, chaining, and burning. Of these processes, the first is the most effective although it is also quite expensive. Root plowing uses a heavy blade drawn by a tracked vehicle to cut roots off below the surface, much like taking the icing off a cake. When combined with stacking and raking, it leaves an area ready for wheeled cultivation. Rome plowing, as previously mentioned, is the use of a heavy disc pulled by a tracked vehicle. This method is usually sixty percent effective when done both ways. In addition, a seeder can be put on the plow to plant at the same time. The roller cutter dices up all brush and trees up to six inches in diameter without destroying existing grasses. Rome plowing is cheaper than root plowing ($35 per acre versus $90 per acre), but this method leaves stumps that will puncture the heaviest of tires.

Chaining involves a heavy ship's chain pulled by two tracked vehicles. This process knocks everything down, but does not work effectively on brush with a trunk diameter of less than two inches since the brush will simply lay down and recover. The process was developed for use in mesquite thickets where trees were the problem. Chaining does not clean the area at all. Burning effectively removes brush without permanently damaging existing grass. Upon evaluation of the pastures, however, the Soil and Water Conservation Service (SWCS) has noted that the grass present was not thick enough to generate the heat necessary to kill the brush. Burning also requires the clearing of fire lanes by bulldozer, which adds to the total cost.

Of all the methods discussed, it is recommended that someone be hired to do the root plowing/stacking method so that the strips can be cultivated as fields. Or, a tracked vehicle is purchased as the prime mover for the ranch and it can be used not only to keep the strips clear, but to cultivate both the strips and the fields. This would depend on the initial investment available, tax situation, and what could be gained from the tracked vehicle versus its higher maintenance costs. The SSCS will pay for 60% of an approved program, and this could well offset the cost of custom root plowing.

Program Implementation. It is suggested that an area be strip-cleared between the midline ridge and the Macho Creek, allowing most of the blackbrush to remain. This blackbrush is a food source to wildlife and does not inhibit the growth of native grasses to the extent that mesquite and whitebrush do. Additionally, some clearing should be done east of the Macho Creek where space allows, but the areas of the back tank and the Macho Creek itself should be left rough. The strips should be zigzagged every 200 yards, and should be 100 yards wide with 100 yards of brush between each strip. This allows enough wildlife habitat/livestock cover while still giving the operator enough room to maneuver. The cleared areas would be managed as one field per pasture and put on the rotation system with the front fields.

Water. Water is, of course, the limiting factor in this area and although the Soil and Water Conservation Service has stated that this is a five-year low, it would be wise to use the present situation as the basis for planning.

Ponds (Stocktanks). Most of the original water capacities of the existing tanks on the ranch have been lost because of silting. Since it is cheaper to build new tanks than to clean old ones, it is proposed that consideration be given to the construction of three new tanks, one tank south of the big tank and two tanks upstream of the back tank. These would be small, two to six acre tanks with deep holes as opposed to the flat design that currently exists on the ranch. Sand in this area prevents any tank from being deeper than twelve feet, but a tank that is ten feet deep and one acre in size would be more enduring than a larger, shallow tank due to the high rate of evaporation in this area. The mud traps on the ranch can be effectively cleaned out using the local drag line since the mud traps are small enough for that machine to reach the center without ever having to move onto the tank bed. This would marginally increase the water capacity, and more importantly, would once again allow the mud traps to protect the larger tanks from silt.

Windmill. Assuming that the lane system of pasture rotation is utilized and dependable water is provided to those pastures, the windmill would be an unnecessary expense. A pumpjack, either motor or tractor driven, could be placed on the well after replacement of the standpipe, cylinder, and leathers. But, as previously stated, this would only be an auxiliary water source and does not play a role in the recommended development of the ranch.

Headquarters Well. This well should be assessed to determine if the water is in fact inferior and what it would cost to return the well to operation. If the well is found to be inferior or prohibitively expensive to rehabilitate, it could be dismissed since the irrigation well, as proposed, will provide all the necessary water.

Irrigation Well. This well should form the basis of the operation. At present, the use of a submersible pump in the well limits its use to the headquarters and pen areas. Also, from experience with the Carrizo waters, submersibles are undependable and will cost the ranch a good deal of money over the years, both in repairs and in down time. It is recommended that the old motor and wellhead be remounted with a 10,000 to 20,000 gallon storage tank built alongside. The tank can be either concrete rings or a poured reservoir, but a metal tank would not last as long.

An interim solution would be to seal the existing flat earthen tank with Bentonite and use it as the reservoir. Troughs on the system would be filled using gravity flow while the headquarters would have the pressure tank to provide the pressure necessary for showers, watering, etc. The addition of a large storage tank would accomplish two things: The large storage tank would provide water to livestock in the event of a well breakdown and it would also improve the quality of the water by allowing it to settle. The size of the storage tank is determined by the requirements for water during a drought should you experience a well breakdown. Using the formula: 200 head at fifteen gallons per day multiplied by seven days equals 21,000 gallons. Fifteen gallons per day per cow is a conservative estimate by local standards.

Troughs. Again assuming that the lane system of grazing management is used, water will need to be provided to seven of the proposed eight pastures as well as to the pens and corrals at the headquarters. Although large troughs would be more convenient to the livestock, troughs of a small diameter, yet deep, would better conserve water. These troughs could be manufactured on the ranch, or could be purchased locally ($150 each). These troughs would be hooked up to a two-inch mainline that is proposed to run down the lane and could be made to serve two pastures at once. Troughs not in use would be shut off to reduce water loss or damage to the system. The pens and corrals would be on the pressure system of the headquarters due to the high water requirements during roundups.

Physical Plant

Headquarters. A major rework of the plumbing system for the headquarters area is required. Old pipes should be replaced with PVC as the pipes become unserviceable. This need not wait on the construction of the proposed water system since the present system originates at the irrigation well. Painting is needed in a number of places and the barn should be painted entirely, including the roof. A kitchen should be added to one of the houses for the manager as well as a screened porch. Yards and parking areas should be observed and the areas should be off-limits to livestock. The cattle guards on the entry road should be cleaned and repaired. Corrals should be added to the barn for horses and kennels might be built on the eastern side of the office building.

Pens. As noted, the ranch does not have any pens. Under the lane system, the pens would be the focal point of all livestock movement thereby reducing the herd's fear of the headquarters area. Figure 12-10 depicts the proposed design of the pens. The pens would be constructed of bull netting on 2″ × 6″ × 10′ boards countersunk into railroad ties. The pens thus constructed would satisfy the needs of the ranch and this particular design can handle 100 head every three hours. The small traps along the mid-fence should be repaired as sick pens and to facilitate the working of the herd. The final chute and the cutting gate should be made of pipe, but the welding can be done on site. The only particular expenses will be the squeeze chute and calf table which are necessary for a smooth, safe processing of livestock. Both can be portable and the expense could be shared with the neighboring ranchers.

Fences. Figure 12-11 shows the proposed new fencing required to establish a lane system of rotation grazing. The cross fences would utilize existing rights-of-way and would be constructed of metal/cedar posts in a ratio of two to one with either heavy posts or welded pipe every 100 yards to aid in stretching/repair. The lane fences, particularly near the pens, would be constructed of netting that was found at the dump near Field #6. This netting is in good condition and will prevent breakouts which often disrupt roundups. All other fences would initially be repaired, tightened, and then put on a rotating replacement schedule in cooperation with the neighbors dependent on the availability of materials and labor.

Wildlife. Pending a cooperative agreement with the neighbors or the game fencing of the ranch, recommendations for management of the game on the ranch are as follows:

Figure 12-10. Proposed Pens

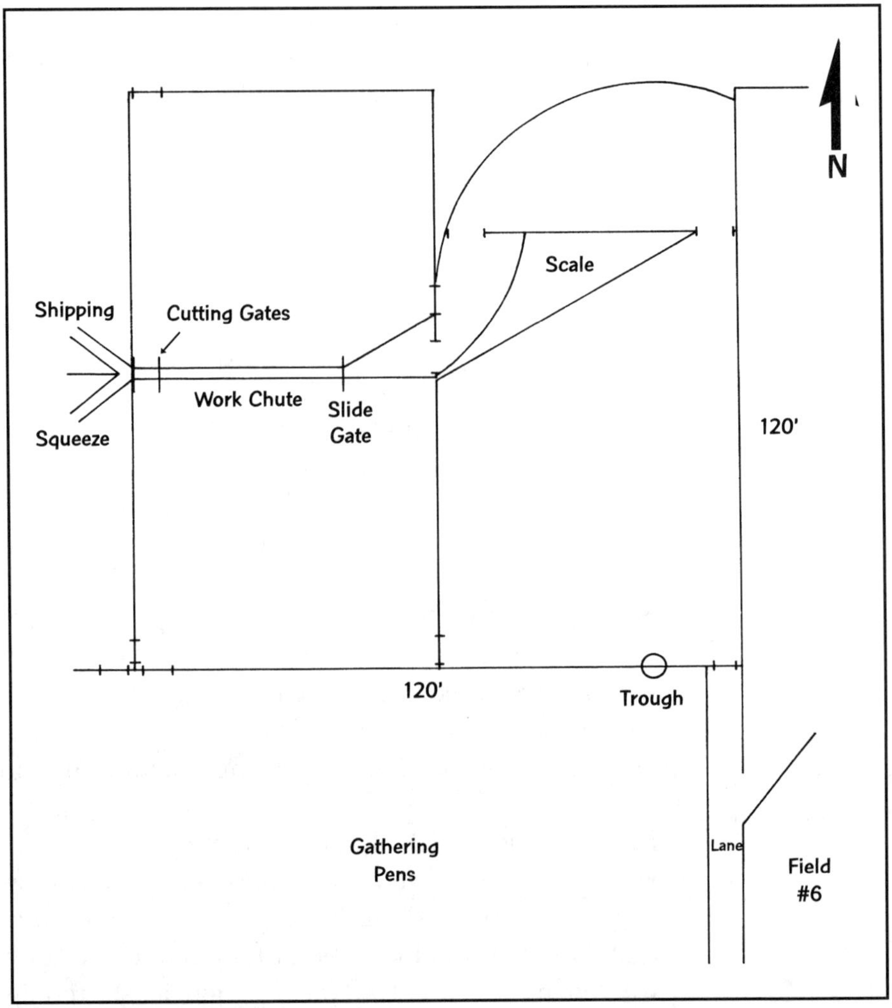

1. **Deer**—During this coming season, the permitted limits as allowed by the Parks and Wildlife Service plus two for ranch use should be harvested. It should also be planned to take all spikes as well as all bucks with four points or less. A policy should be implemented to shoot any buck with narrow or deformed antlers, and all deer that appear to be sick or injured. Good bucks should not be taken under any circumstances for the first six years and this policy should be emphasized to all hunters. At this time, feeders are not recommended since the free movement across

Figure 12-11. New Fences, Water Stations, and Strip Clearing

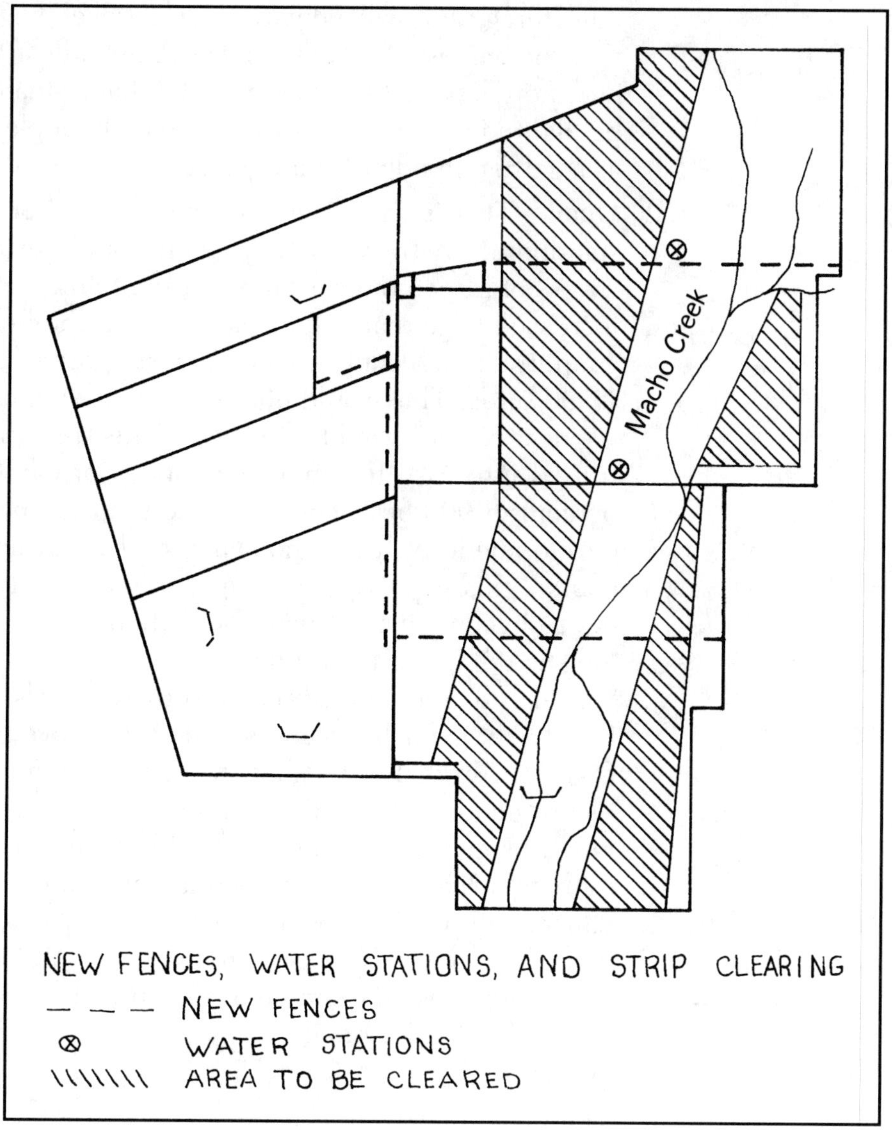

the fences from ranches not properly managed will negate any attempts made at improving the local deer population. Mineral blocks at the time of antler development are recommended for deer from an economic standpoint. Leases for deer hunting could accomplish everything as described above, but it is the rancher's decision whether to lease or depend on guest hunters.

A screened game room attached to the office would facilitate storage and will aid with the recordkeeping process. To further the harvest of does as well as to provide an indication of the deer population, it is recommended that a small area of Field #6 be planted with oats, not as a source of supplemental feed, but as a gathering point for area deer.

2. **Quail**—This is the ranch's strongest suit from a management standpoint. With water, the ranch could sustain seventy plus coveys of fifteen birds each. With supplemental feeding and the increased edge that will be gained from the brush clearing program, the ranch could conceivably support a total population of 1,500 birds. This would not only give the rancher all the hunting desired, but would give the rancher a surplus that could be turned into cash for improvements. A prudent harvest of sixty percent or 900 birds could be taken in the final stages of ranch development by the landowner or by paying customers. Of course, this stage is five to eight years away, time for new coveys to become established. Until then, quail can be hunted as long as the harvest is prudent.

   It is recommended that 175 to 200 birds be harvested this fall. And it is further recommended that coveys that flush fewer than four birds not be shot. More important though, it is also recommended that water stations be established across the ranch using the design developed in California for desert environments. These water stations are cheap, require little maintenance, and provide the birds with enough water not only to live in an area, but to increase through successful hatchings. Figure 12-11 provides an example of the spread of water/feed stations recommended for quail.

3. **Dove**—Because of the migratory nature of dove, these birds cannot be managed as such. Dove can, however, be attracted to areas with food and water. It is recommended that the natural flyway that crosses the ranch be capitalized upon and that areas near existing tanks be planted with grain to attract/stop passing dove. In Charlotte, a number of flyers were found from people who want dove leases. Since the ranch has an unmanageable resource, which is renewable on a daily basis, it is fur-

ther recommended that beginning this fall, "dove only" hunting rights for the ranch be leased. The contracts can be either "rabbit rights" such that the hunters bring their own shelter, or for a substantially higher price, they could use one of the houses in the compound.

4. **Fish**—At present, this is a low priority on most ranches. It has been predicted, however, that by the year 1999, the average American will consume twice as much poultry as beef and the same amount (fifty pounds per year) of fish that consumers now eat in poultry. At this time, it is not recommended that a poultry program be initiated both from an initial investment and an experience viewpoint. But a fish program can be recommended as an economical, easy, and quick way to increase profit. The new tanks should be constructed from the first with this in mind and existing tanks should be cleaned and managed for commercial catfish production. There are now a few such operations in Atascosa County with a growing market in San Antonio.
5. **Miscellaneous Wildlife**—Javelinas, feral hogs, and bobcats are not protected in Texas. We believe that, with the exception of bobcats, these too could be harvested at a profit to the ranch. We are proposing not a supermarket for wildlife, but rather a diversification approach to increasing ranch income while protecting the habitat for all species. Lease hunting cannot be overlooked as a source of income for the ranch and it is believed that there is substantial surplus game, exceeding that needed for guest hunting.

Addendum: Why a Lane System of Rotation Grazing Is Recommended

Throughout this survey, the development recommendations have been based on the premise that an eight-pasture rotation grazing system be utilized. Rotation grazing involves grazing each pasture for a short period of time, e.g., seven days, then moving the stock to the next pasture in regular succession. This system is based on the assumption that animals in large numbers make a more uniform use of the forage, and that a rest from grazing is beneficial to the plant.

Recovery time for coastal is twenty days and for Kleingrass is twenty-four to twenty-seven days. This rotation actually increases the yield of the plant even though it must support a greater number of animals during the shorter time during which it is grazed. Proper rotation grazing results in more uniform utilization since a large number of animals in small pastures are forced to move over the entire area and use the available forage more uniformly. Trampling is reduced because animals in smaller areas where food is more abundant need to travel less and are in place for a shorter period of time. One application of this system is the Merrill four-pasture system whereby four equal pastures are made out of a tract. Three are grazed at any one time and the fourth has three months deferment out of twelve. In the case of this ranch, the Merrill system would work except that the ranch is divided between fields and rangeland and the fields could not support the herd for that period of time without reducing productivity. A second variation is called the cell system (see Figs. 12-12 and 12-13). It differs from the recommended plan only in that the water is dispersed from a central point. The advantages to this are as follows: The herd must come to the pens on a daily basis if the pastures have no other water, and the gathering of the herd is simplified by the funnel effect of the fencing. Drawbacks to this system in the case of this ranch are as follows: A fence would have to be strung through the brush while the lane system uses existing rights of way, and the use of Field #6, one of the cleanest and easiest to cultivate on the ranch, would be lost.

Additionally, although supplying water to the farthest pastures would require some effort, with eight pastures, one or two fields can be deferred while under cultivation without overgrazing any one portion of the ranch. Then too, movement of the herd on a weekly basis, even if only across one fence, will help tame the animals and acquaint them not only with the pens, but with a person on horseback. It has been shown that after a period of three months or less, the herd will expect to move on the last day and will not need to be enticed. Certainly, movement is determined by forage availability, not the calendar, and a certain eye for forage quality is required.

Figure 12-12. Lanes and Water Trough Placement with My Fence Proposal

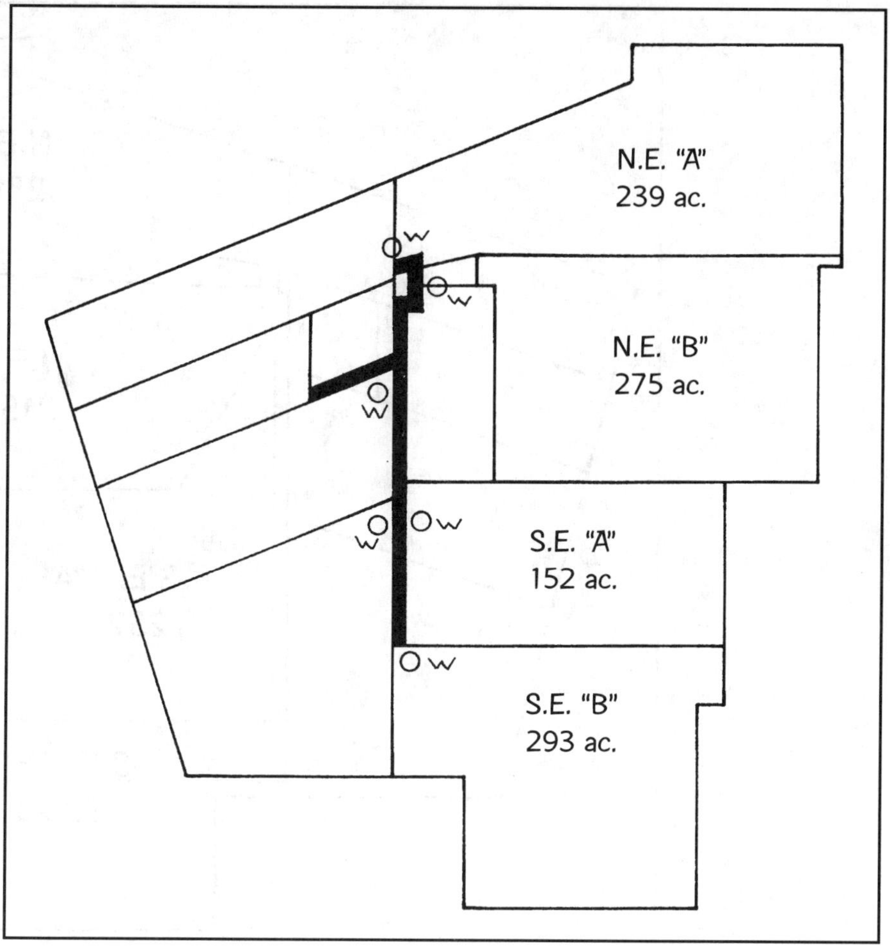

Figure 12-13. Fences as Proposed by the Soil and Water Conservation Service

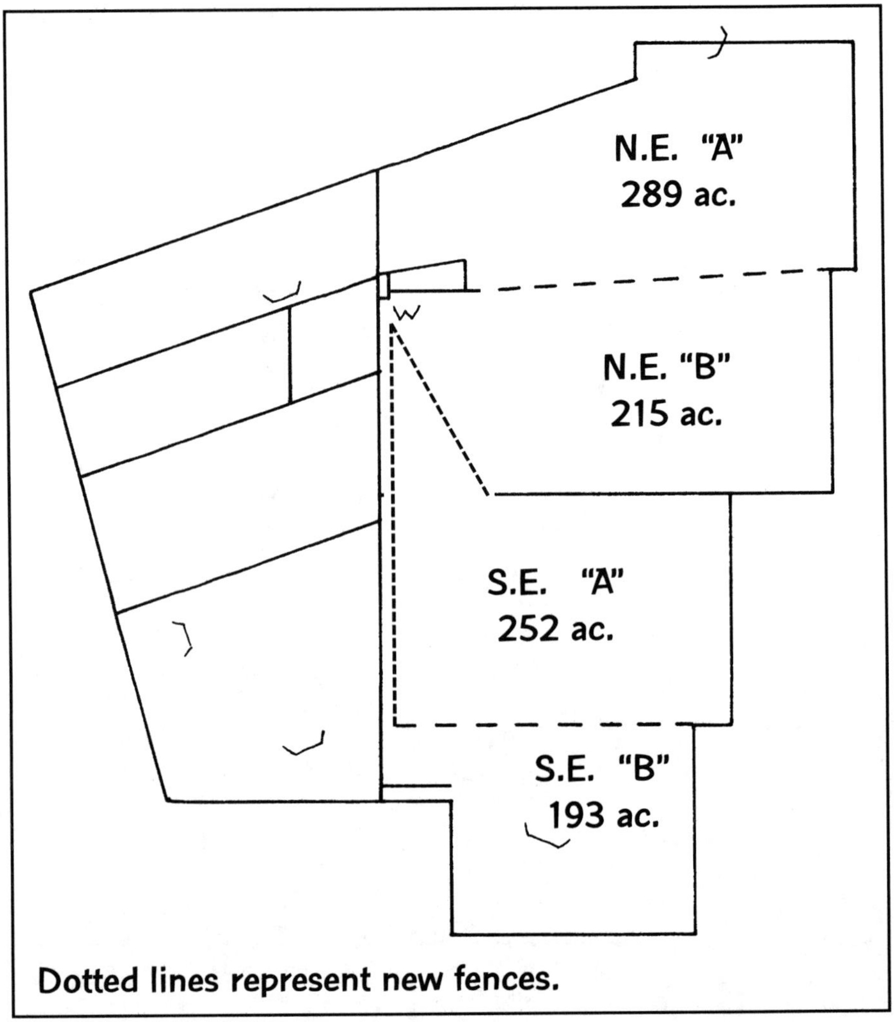

13
Wildlife Management for Rural Real Estate Professionals

In my experience, today's buyers are most often successful professionals, usually at least one generation away from the land. They are looking to recapture the pastoral life they enjoyed with their parents or heard about at family gatherings. They once had rural land in the family, but it was sold or lost over the years. For the most part, they want the land to produce, usually with a "hobby" herd of cattle, perhaps some light farming, usually hay, and horses for the kids.

Wildlife is always a concern for these buyers. In parcel sizes over 250 acres, most are hunters and want their children to learn the joys of hunting, the responsibilities of the kill. They invariably want to actively manage wildlife at any rate, even if only to improve viewing opportunities for their family and friends. As mentioned, wildlife always adds to the value of rural real estate and is often the prime value of a piece of land.

In this chapter we will review some basic wildlife management principles. While these principles are specific to the Southwest, they can be applied across the country with variations for differences in climate, topography, and vegetation. The purpose of the chapter is not to make the practitioner a biologist, but rather to provide general management ideas that will prove useful when dealing with the buyer. Professional farmers and cattlemen rarely need input in the buying decision. Recreational buyers, those "returning to the land," on the other hand, are often ignorant of natural resource management techniques. The practitioner can be of great assistance in this area.

MANAGEMENT FOR CONSUMPTIVE USE OF WILDLIFE

Whitetail Deer

From the eastern seaboard to the desert southwest, the principal game species is the Whitetail deer. Once reduced to fractions of its original numbers, the Whitetail population has recovered with the help of sound management to number once again in the millions. Hunting the Whitetail provides innumerable small towns with critical tourist dollars, gives parents a chance to spend time with the kids in the outdoors, supports wildlife management at the federal and state levels through taxes on guns, ammunition, and supplies, and, incidentally, sells a lot of real estate.

The Whitetail is very adaptable in its living requirements. It occurs in deep forest as well as open prairie, usually seeking what is known as the "edge" for its preferred habitat. Edge is just that, the transition zone between grassland and brush, between brush and forest, forest and meadow. It allows the deer to forage for preferred foods within easy running distance to safety. The Whitetail has a home range of around 300 acres for a doe and 500 acres for a buck, although this is extremely variable. They mate or rut in the fall, when day length shortens. This allows them to birth in spring when foods are fresh and green and most nutritious for the doe.

Whitetail fawn in hiding

Preferred foods are forbs, or weeds, over most of the year, supplemented with the tips of brush plants or browse. Deer rarely eat grass; perhaps only 2% of their diet is grass. This is because, except when it is very young, grass is hard to digest and low in nutrients. They consume mast or seeds when available; acorns and mesquite beans are very important to deer where those fruits occur. Deer require water to live, although they can survive on the water contained in their food if the climate is not too dry. To reproduce, however, free water is a must and deer will travel long distances in drought to reach drinking water.

Management for Whitetail deer is a combination of habitat manipulation and population control. Since they are creatures of edge, oftentimes land managers will clear strips or blocks in dense brush/forest and can remove up to 50% of the vegetation without adversely impacting deer numbers. This clearing increases forb (weed) production, yet retains covered access for the animals throughout their range. Those wishing to hunt or view deer often plant oats, rye, or clover as a cool season food supplement. Recent developments indicate that warm season food plots using Lab-lab, white milo, or peas are beneficial as well. Stress periods for Whitetail deer are in late winter and late summer. Of the two, the former is most important as the doe is carrying the fawn at that time.

Population management is based on a census of the population. There are several census techniques, all of which provide an estimate of the herds' numbers. The Hahn Cruise method uses repetitive walks along a set trail, counting deer seen and the distance from the trail. Taking the length of the trail (usually a minimum of two miles) times the average distance to deer seen gives the manager the area surveyed. Taking that area and applying it to the number of deer seen, the manager can determine what animals, if any, need to be harvested. Another method involves the use of a spotlight and a vehicle. Similar to the Hahn Cruise, the spotlight count uses repetitive trips along a set route at night. Spotters record deer seen and estimate distances from the vehicle. This gives the area surveyed and some idea of the population. Last, in areas of light or short brush, helicopters are used to survey deer numbers. By flying in parallel transects or lines across a tract of land, a certain percentage of the deer herd can be seen. While it

Prepared food plot

takes no experience to see and count deer using this method, some experience is useful not only in determining the speed of the helicopter, the width of the transects, and the best height to fly, but also in determining what percentage of the herd was actually seen. Studies indicate that between 50% and 80% of the deer are seen in this manner, quite a range in accuracy if not addressed.

Once estimates are made of the population, the manager must determine what the sex ratio is, what the density is, and what percentage of the does had fawns. The sex ratio is important primarily to those operations managing for bucks. In such operations, one buck per two does is the norm, even though one buck can breed or "cover" up to twelve does successfully. The reason for the emphasis on bucks is obvious; the more bucks, the more opportunity for a "trophy" to mature and produce his best antlers. In fact, some operations strive for a one to one ratio to maximize the number of males. This, however, often leads to broken antlers and physically worn bucks as the competition for sex increases.

Density is the most important factor in Whitetail deer management. It is density that determines whether each individual will have enough nutrition to maximize its genetic potential or merely survive. Left unmanaged, Whitetail deer will overpopulate their habitat, not only precipitating a die-off, but destroying the habitat for years to come. Ideal density is related to carrying capacity and thus varies from place to place, day to day, and season to season. Density is given in acres per adult. The wise practitioner will seek out wildlife professionals to learn objective or target densities in a given area.

The last indexes to be obtained from a deer survey is the fawn crop, or reproductive ratio. It is given as a ratio of does to fawns and is a good indicator of general herd health. A young doe, given adequate nutrition, will have a fawn every year and twins are common in healthy, managed herds. Therefore, a fawn crop of 100% indicates that every doe had one fawn, on average, thus the herd is replicating itself every two years. It is important to note the time the survey was taken when considering fawn crops. The later in the fall the count is made, the greater impact fawn mortality has had on the numbers. Late fall counts may indicate a fawn crop of 60% when in fact every doe had a fawn, but many were lost to predators.

A growing management technique involves the use of high fences to enclose deer herds. This allows for intensive management by reducing the normal ingress/egress of deer. Because of the reaction of the enclosed herd to intensive management, fawn crops often approach 150% and removal of surplus deer becomes a logistical problem. Any landowner or buyer is cautioned to carefully research deer management under high fence. The problems may well outweigh the benefits.

Another popular management technique is supplementation. This effort assumes that nutrition quality or quantity is limiting the herd from reaching its genetic potential. True supplementation, other than the food plots mentioned above, involves prepared feed or minerals for deer. Corn is not usually considered a supplement but rather a bait as corn is not nutritionally balanced. In any supplementation program, it should be remembered that artificially raising the carrying capacity increases the managers' responsibility. Supplementation should always be a part of a total management program, not a quick fix for poor management. Remember, too,

High fence used in management of Whitetail deer

that stopping a supplemental feeding program after raising the carrying capacity will be detrimental to the natural habitat and potentially disastrous to the population. Even planted foods for deer should be managed to provide food over all stress periods, not just during hunting season.

Bobwhite Quail

Once limited to plantation-style hunts in the deep South, the Bobwhite quail is quickly becoming the preferred game species for many hunters. The bird offers great sport as it is quick, unpredictable, and quite tasty. Unfortunately, it suffers from extreme fluctuations in numbers oftentimes for reasons as yet unknown. The birds are social creatures, spending most of the year in coveys ideally numbering 12–14. This is for security as they roost in the open, in a circle facing outward. A covey of fewer than 12 reduces the number of sentinels, increasing the workload for all birds and allowing for increased predation. The birds pair off when green-up occurs in spring, the hen making a nest in tall grass when available, using almost anything when it is not. Unlike Whitetail deer, the male assists in the incubation and care of the young, and can even take over the care entirely, allowing the hen to nest again if conditions are right.

Bobwhite quail are creatures of early succession, meaning they thrive immediately following disturbance. They eat a wide variety of foods including seeds, insects, green forbs, and fruits or mast. The young require two weeks of insects at first, which is sometimes limiting in times of drought. At one time, it was thought that in more arid climates, the lack of free, or standing, water was limiting quail numbers. Recently, it appears that Bobwhite quail are able to obtain enough water from their food to successfully reproduce; however, supplemental water in the form of sprinklers may, through temperature reduction, increase Bobwhite nesting success by increasing chick survival.

As Bobwhite quail begin dying the day they are born, there is rarely a problem of overpopulation. For this reason, census taking is not normally a part of management. Instead, the landowner/manager strives to increase numbers, allowing for a heavy harvest

Wildlife water trough

Sprinkler used in quail management during hot summer months

while preserving adequate numbers for nesting. Habitat manipulation that increases forb production while reducing woody cover is best for Bobwhites. Up to 80% of the woody vegetation may be removed without impacting quail numbers adversely. Logging, chaining, "stacking," discing, and burning all can be used to increase Bobwhite numbers and distribution. Chemical treatments can reduce brush/tree cover and are often followed with burns to further reduce grass and increase weeds.

Supplementation in Bobwhite quail management is more a matter of economics than carrying capacity as food is rarely the limiting factor in a natural population. As already mentioned, Bobwhites are dying from many factors from day one. Adding food can be done to draw them to the hunter, but often as not, will not increase survival. In addition, drawing the birds to feeders increases their susceptibility to predation and disease, offsetting any benefits supplemental food may have. When provided, supplemental food should be broadcast over large areas, into good cover. This will allow the birds to utilize the food without undue exposure and without concentrating them.

Burning improves quail habitat when properly applied

An example of a simple quail feeder

More so than deer, Bobwhite quail do better on some soil types than others. Because sand is so permeable, light rainfall on sand will produce weeds more readily than on a heavier clay soil. A succession of weed crops may not provide more food over time than one major crop, but increased diversity and availability over time increases the birds' survival chances.

Doves

Mourning doves are migratory birds that provide hunting opportunities throughout their range. White-winged doves are more stationary, colonizing mature trees in towns in place of the dense brush along the Rio Grande that once served as their nesting cover. Both species offer excellent hunting opportunities in their ranges, often with minimal management.

The key to managing for doves is to provide one or both of their primary requirements: food and water. The birds feed twice a day and water twice a day. Establishing surface water with bare soil around the edges, to facilitate landing, will attract doves as will planting either small grains or flowers. Delaying the discing of a field after harvest of a small grain crop will draw birds as will leaving strips unharvested. Supplemental feed from a bag is illegal, while planting is not, so plan any food management for doves well ahead of the season.

Food plot for doves, quail, and turkey

Turkey

The wild turkey is another success story of wildlife management. Hunted to near extinction, it has now been reintroduced to all of its former range and is expanding into areas where it previously did not occur. In most areas, populations are strong enough to support hunting pressure and many states have adopted two hunting seasons, one concurrent with deer season and another in the spring. Spring hunting involves luring turkeys into range with imitations on the mating call and is a challenging, growing sport.

Turkeys are extremely cautious birds and perhaps the best management for them is to leave them alone. Their life requirements are trees to roost in, tall vegetation to nest in, food, and water. Food in the form of corn or other small grains is readily taken from feeders and can be used to hold a flock in an area. To reduce loss of feed to other animals, turkey feeders are often placed on platforms which, when skirted with sheet metal, also serve to protect turkeys from raccoons and bobcats.

Turkeys require tall vegetation for nesting, and will not remain in an area overgrazed by livestock. Burning promotes insects, a fa-

Turkey feeder on platform

vorite food in the spring and summer, as will discing, by increasing weeds and flowers that attract insects. Turkeys roost in large trees, preferring live oak and hackberry in South Texas. Where there are no large trees, artificial roosts can be built, using telephone poles implanted vertically and connected on top with 1½-inch pipe. Sheathing the poles with sheet metal again discourages uninvited guests.

Fish

Perhaps the easiest consumptive asset to manage is fish which, once started, require only to be culled regularly to remain healthy. Even the smallest tract can support a pond or stocktank, particularly when it can be supplemented by a well. Fishing holes need not be large, but should be fairly deep. Quality fishing can be maintained in a one-acre pond if it is deep enough. The need for depth is due to the temperature layers formed in water, called thermoclines. Depth allows the thermocline near the bottom to remain cool in hot weather; cooler water holds more oxygen, thus prevent-

Digging a new pond

ing loss of fish to oxygen depletion. As a general rule, ponds should be at least seven feet deep to protect fish from hot weather.

Adding tires, brush, or old equipment adds structure to the bottom, increasing cover for forage fish and improving fishing quality. Specially formulated food can be added to a pond to increase carrying capacity but be careful as always to remove surplus individuals to prevent overpopulation, stunting, and die-offs.

Proper management of a fish population is always easier if you know what you are managing. For this reason, and because so many ponds are choked with trash fish, many new landowners remove all fish and restock. This can be accomplished in any of three ways: electro-shocking, which stuns the fish so they can then be netted; seining, which works on smaller bodies of water; and rotenone, a plant derivative that coats the gills, bringing the fish to the surface, allowing them to be netted. Of the three, electro-shocking is best as it allows the manager to decide if total removal is necessary without harming the fish.

Once a pond or lake is cleaned of all fish, restocking is accomplished in two stages. The first is the stocking of a forage fish

Adding structure improves fishing quality

Trash fish must go for management to be successful

species, usually a variety of sunfish commonly known as perch. These can be accompanied by a minnow or a shad species. Then, after the forage fish have had time to establish themselves and reproduce, a predator fish, almost always bass of some sort, are stocked. This completes the food chain and is pretty much self-supporting. Catfish can be added to any pond without adversely affecting the overall quality of fishing. This adds another type of fishing that may be productive when the bass are not biting.

Problems arise if the pond does not get fished often enough. This results in overpopulation of both forage fish and predator fish and an overall reduction in fish size.

Aquatic vegetation is not a requirement of good fish management although some may be left as structure. Several chemicals are available to reduce vegetative growth in ponds and to open up waterways. Surprisingly, these chemicals do not harm fish and livestock can continue to use the pond for drinking. One cautionary note: Do not treat more than half of a pond's total surface at one time as the death of large amounts of aquatic vegetation will sharply reduce oxygen levels in the water.

Good management equals happy fishing enthusiasts

MISCELLANEOUS CONSUMPTIVE WILDLIFE MANAGEMENT TIPS

The foregoing was a treatment of the major game species in the Southwest. Other regions of the country will have devotees to other species and management techniques. Limitations in space and experience restrict the number of species discussed here. Below are some random tips that may come in handy.

Ducks and Geese

Ducks and geese are migratory and limited in the extent of their flyways. For this reason, there is little the landowner can do if not on a flyway. If the land presently has ducks and geese for some period of time, adding to the availability of food and the accessibility of water will increase their numbers and duration of stay.

With planning, a flat area surrounding a pond can be modified into a duck marsh by raising a small levee. The area enclosed by the levee can be planted in small grains such as Japanese millet, then flooded when ripe to attract and hold ducks. Geese are attracted to fields and leaving grain in the field after cutting will draw geese into the area. The same applies to sandhill cranes, who have similar habits to geese.

Exotics

Management of exotics is similar to that of livestock with one major difference. Exotics can and will out-compete native wildlife for food and will, if not carefully monitored, destroy native habitat. So many times, buyers are looking for a quick buck and stock their new ranch with exotics. After several years, the market for whichever exotic they have dies down, leaving them with mostly females that no one wants. Feed bills go up, income goes down, and the habitat and native wildlife suffer. Exotics are legally considered to be livestock and should be managed as such. Learn what animal unit equivalent each exotic species is, what it eats, and what density the land can support. Include this density when determining stocking rates of domestic livestock and native wildlife, and stick to the optimum stocking density. If the market for the exotic falls, or reproduction exceeds harvest quotas, remove the surplus by whatever means necessary, but get them off the land. Destroy the habitat and you not only cannot carry any animals, you have a poor property for resale.

MANAGEMENT FOR NONCONSUMPTIVE USE OF WILDLIFE

With the urbanization of America, a growing industry has arisen around the "outdoor experience." This industry serves urban dwellers who want the outdoor experience away from state or federal parks, away from the crowds, and in a more natural setting. Any rural tract has potential for nonconsumptive wildlife income and simple management can increase this potential.

The chief nonconsumptive use for wildlife in terms of numbers and dollars spent is birdwatching. Literally millions of Americans each year go afield armed with binoculars and bird books. As a rule, they are unobtrusive, neat, friendly guests. Adding surface water, letting brush grow back, and planting small grains will increase bird numbers and diversity for birdwatching. Specialized improvements, such as the duck marsh described above, will greatly increase income from this activity which can be sandwiched in with hunting to maximize income. Another improvement to aid birdwatchers is the construction of blinds near feeders—again, something that serves a dual purpose where hunting is allowed. Many hunters graduate to cameras with age, either because they have taken their trophy, or because they want to have a taped record of what animals they've seen. A blind is a blind, and a good one can easily serve dual roles.

Hiking, camping, and boating are all nonconsumptive uses of wildlife and provide the outdoor experience. Tracts with permanent water, rugged topography, and dense vegetation are particularly attractive to the camper. Rural properties can be adapted for these nonconsumptive uses by simply improving access with trails, designating parking areas, making water available, and establishing simple sanitary facilities.

Conservation of endangered ecosystems is another nonconsumptive use of wildlife. This category includes wild plant life, insects, geology, and historical sites. Large areas near Austin, Texas

Perfect habitat for campers, hikers, and birdwatchers

are being set aside for the Black-capped vireo and the Golden-cheeked warbler. Buyers for this type of land intend to preserve the critical habitat for these species. Across the continent, landowners are being encouraged and assisted in setting up preserves for butterflies, going so far as to plant specific wildflowers to feed them during their migration. The hummingbird migration along the Gulf Coast is a well-publicized event, with tracts of land specifically set aside for those birds to rest on.

Finally, there is growing interest in the use of skulls, antlers, and dried flowers in artwork. Simply allowing people access to collect these items can generate income for the landowner with no overhead, no facilities, and no time required on his or her part.

There is always concern among buyers about liability. The practitioner should be aware that large judgments have been awarded against landowners for parties injured on their land. However, there are defenses against lawsuits and insurance coverage for just such uses. Potential litigation should not prevent properly prepared new landowners from enjoying their property or deriving income from it through marketing of the wildlife assets.

An old satellite dish recycled as a wildlife water hole

In arid areas, rainfall is gathered and stored for wildlife

Index